COTTAGE ECONOMY

WILLIAM COBBETT

COTTAGE ECONOMY

CONTAINING

Information relative to the brewing of BEER, making
of BREAD, keeping of COWS, PIGS, BEES, EWES,
GOATS, POULTRY, and RABBITS, and relative to other
matters deemed useful in the conducting of the
affairs of a Labourer's Family; to which are added,
Instructions relative to the selecting, the cutting
and the bleaching of the Plants of English GRASS
and GRAIN, for the purpose of making HATS and
BONNETS; and also Instructions for erecting and
using Ice-Houses after the Virginian manner.

WITH A PREFACE BY
G. K. CHESTERTON

Oxford   New York   Toronto   Melbourne
OXFORD UNIVERSITY PRESS
1979

Oxford University Press, Walton Street, Oxford OX2 6DP

OXFORD LONDON GLASGOW
NEW YORK TORONTO MELBOURNE WELLINGTON
KUALA LUMPUR SINGAPORE JAKARTA HONG KONG TOKYO
DELHI BOMBAY CALCUTTA MADRAS KARACHI
IBADAN NAIROBI DAR ES SALAAM CAPE TOWN

First published in book form by C. Clement 1822
Seventeenth edition published by Anne Cobbett 1850
Reprinted with introduction by G. K. Chesterton 1916
First published as an Oxford University Press paperback 1979
at the suggestion of R. H. Hardy

British Library Cataloguing in Publication Data

Cobbett, William
    Cottage economy. – 17th ed.
    1. Livestock – Great Britain   2. Home economics,
    Rural – Great Britain
    I. Title
    639.08′0941   SF71   78–40898

    ISBN 0–19–281270–X

*Printed in Great Britain by Cox & Wyman Ltd
London, Fakenham and Reading*

PUBLISHER'S NOTE ON THE TEXT

Cottage Economy was first published by C. Clement in 1821–2 in seven separate numbers, reissued together in book form in 1822. A series of new editions followed, incorporating a succession of revisions and enlargements, beginning with the addition of No. 8 in the second edition (1824). The present text is that of the latest, seventeenth edition, published by Cobbett's wife Anne in 1850. G. K. Chesterton's Preface was added in a reprint of 1916 published by Douglas Pepler, and was also included in a reprint of 1926 published by Peter Davies, from which the present edition is photolithographically reproduced.

1979

PREFACE

WILLIAM COBBETT *is the noblest English example
of the noble calling of the agitator. The term has come
to have a bad sense by a continual reference to cases,
some of them true but more of them mythical, in which
it has been connected with artificial programmes and
with private aims. The truer element refers to a few
quacks who have flourished nostrums which were merely
novelties. The false is part of a snobbish fairy tale, by
which a demagogue was needed to tell a starving man
that hunger hurt him, and another to explain to some
prostrate person that a policeman had knocked him
down. But Cobbett had two clear grounds of defence
against the charge of cheap tub-thumping, in those
days when he sent a fiery cross through South England,
which is perhaps the next thing to setting the Thames
on fire. His first defence is that his type of demagogy
had all the dangers of isolation. He was far too popular
to be fashionable. He spoke for those innumerable who
are also inarticulate; and those he sought to help were
impotent to help him. He was not paid by the poor to
champion their cause: for it is a singular fact, undis-
covered by most of our doctors of sociology, that wealth
is to be obtained from the wealthy.*

*The second fact that cleared Cobbett of the charge
of quackery was that his nostrums were not novelties,
but very much the reverse. To use the language of a*

*religious world which he furiously detested, he was a
revivalist. Despite the other connections of the phrase,
the real agitator has to be a revivalist : he has to appeal
to what remains of a memory, or at least of a legend.
What Cobbett attempted to revive was something which
almost all political schools in his time especially despised, that is especially misunderstood : it was really
mediæval England. For the more immediate purpose
of politics, it was rural England. But it was not a
Byronic repose in a rural barbarism; it was a quite
business-like belief in the possibility, or rather the
necessity, of a rural civilisation. He believed that
agricultural labour could pay; he even entertained
the Quixotic fancy that it might pay the agricultural
labourer. But that this might come about, he felt it as
primarily necessary that the labourer should not be a
serf, and even as little as possible a mere tenant. For
the purposes of the present introduction, the most important fact is that he saw the cottager as master of his
cottage; and had the historical instinct to grasp the
great virtues that go with such a small estate. Through
all his days he thirsted after freedom. And he understood something that can only accompany freedom,
property; and something that can only come with
property—thrift.*

*What distinguishes Cobbett from most rural ideal-
ists, such as Ruskin, is that he was a realist as well.
Like Ruskin, and long before Ruskin, he denounced
the eating up of England by factories and industrial
towns. He must have the more credit because he had
not, like Ruskin, the advantage of living when the
terrible transformation was almost complete; when it*

was well within sight of its present congestion and collapse. He defied industrialism when it was, if not exactly young and beautiful, at least young and hopeful. But what distinguishes him, as I say, is the practical upshot of his Arcadianism. This can be seen if we compare him with Ruskin even upon Ruskin's own most sacred ground. With no æsthetic culture and nothing of what men would now call a mystical temper, he nevertheless, by his own independent imagination, realised as fully as Ruskin did the overpowering historic importance of the old churches of England. But even here he shows that note of practicality which is also the note of hope. While Ruskin considered how many carvings could be found in a church, Cobbett always considered how many people could be seated in it. An unamiable critic might say that Ruskin knew everything about the building of a church except what it was built for. This would be exaggerative; but it is really relevant to note that Cobbett, in that utterly un-Christian epoch, did understand what it was built for; for it is the same pointed and fruitful attitude that he occupies towards other things, especially towards that thrift of the cottager which is the matter of this book. Ruskin could be trusted to tell his pupils how they should labour with paint or pencil to reproduce every vein and tint upon a cabbage leaf. But few would have trusted Ruskin with the cooking of the cabbage.

" Cottage Economy " is a book which belongs entirely to this practical and even materialistic side of Cobbett's campaign. Its value, though of the most valid kind, is not of the sort for which it is possible to plead in pen and ink. A cookery book can scarcely be a basis of

controversy, though it may be of combat; and the proof of the pudding is in the eating. This is merely the commissariat of his revolutionary army; and, like a good general, he paid a great deal of attention to it. But scattered even through these pages, as through all the pages he wrote upon any subject, there are numerous lively passages which give us glimpses of his philosophy. It can hardly be missed in the case of those two grand survivals of a more Christian England, bacon and beer; but it is quite equally apparent in the study of so small a matter as mustard. I do not profess to know by what process Cobbett discovered that the mustard bought in shops is adulterated, or even relatively poisonous. But it is a perfectly sound criticism on the anonymous tyrannies of trade that we have no possible means of knowing that it is not. The mustard seed that Cobbett advised the cottager to grow in his cottage garden is in this matter as symbolical as the similar seed in the parable. Such seed if sown by the genuine English peasant may yet in truth grow into a great tree; and if we had faith as a grain of mustard seed we could indeed cast all our mountains of oppression into the sea. For a hundred years after Cobbett's forlorn hope we are confronted again by Cobbett's question. We must go back to freedom or forward to slavery. The free man of England, where he still exists, will doubtless find it a colossal enterprise to unwind the coil of three centuries. It is very right that he should consider the danger and pain and heart-rending complication involved in unwinding that coil. But it is also proper that he should consider the alternative; and the alternative is being strangled.   G. K. Chesterton.

CONTENTS

The block on page 185 has been engraved on wood by Eric Gill after the original. The scale of feet given at the bottom has been corrected.

COTTAGE ECONOMY

No. I

INTRODUCTION

TO THE LABOURING CLASSES OF THIS KINGDOM

———

1. THROUGHOUT this little work, I shall number the Paragraphs, in order to be able, at some stages of the work, to refer, with the more facility, to parts that have gone before. The last Number will contain an Index, by the means of which the several matters may be turned to without loss of time; for, when economy is the subject, time is a thing which ought by no means to be overlooked.

2. The word Economy, like a great many others, has, in its application, been very much abused. It is generally used as if it meant parsimony, stinginess, or niggardliness; and, at best, merely the refraining from expending money. Hence misers and close-fisted men disguise their propensity and conduct under the name of economy; whereas the most liberal disposition, a disposition precisely the contrary of that of the miser, is perfectly consistent with economy.

3. ECONOMY means management, and nothing more; and it is generally applied to the affairs of a

house and family, which affairs are an object of the greatest importance, whether as relating to individuals or to a nation. A nation is made powerful and to be honoured in the world, not so much by the number of its people as by the ability and character of that people; and the ability and character of a people depend, in a great measure, upon the economy of the several families, which, all taken together, make up the nation. There never yet was, and never will be, a nation permanently great, consisting, for the greater part, of wretched and miserable families.

4. In every view of the matter, therefore, it is desirable, that the families of which a nation consists should be happily off; and as this depends, in a great degree, upon the management of their concerns, the present work is intended to convey, to the families of the labouring classes in particular, such information as I think may be useful with regard to that management.

5. I lay it down as a maxim, that for a family to be happy, they must be well supplied with food and raiment. It is a sorry effort that people make to persuade others, or to persuade themselves, that they can be happy in a state of want of the necessaries of life. The doctrines which fanaticism preaches, and which teach men to be content with poverty, have a very pernicious tendency, and are calculated to favour tyrants by giving them passive slaves. To live well, to enjoy all things that make life pleasant, is the right of every man who constantly uses his strength judiciously and lawfully. It is to blaspheme God to suppose that he created men to

be miserable, to hunger, thirst, and perish with cold, in the midst of that abundance which is the fruit of their own labour. Instead, therefore, of applauding " happy poverty," which applause is so much the fashion of the present day, I despise the man that is poor and contented; for such content is a certain proof of a base disposition, a disposition which is the enemy of all industry, all exertion, all love of independence.

6. Let it be understood, however, that, by poverty, I mean real want, a real insufficiency of the food and raiment and lodging necessary to health and decency; and not that imaginary poverty, of which some persons complain. The man who, by his own and his family's labour, can provide a sufficiency of food and raiment, and a comfortable dwelling-place, is not a poor man. There must be different ranks and degrees in every civil society, and, indeed, so it is even amongst the savage tribes. There must be different degrees of wealth; some must have more than others; and the richest must be a great deal richer than the least rich. But it is necessary to the very existence of a people, that nine out of ten should live wholly by the sweat of their brow; and, is it not degrading to human nature, that all the nine-tenths should be called poor; and, what is still worse, call themselves poor, and be contented in that degraded state?

7. The laws, the economy, or management, of a state may be such as to render it impossible for the labourer, however skilful and industrious, to maintain his family in health and decency; and such has,

for many years past, been the management of the affairs of this once truly great and happy land. A system of paper money, the effect of which was to take from the labourer the half of his earnings, was what no industry and care could make head against. I do not pretend that this system was adopted by design. But, no matter for the cause; such was the effect.

8. Better times, however, are approaching. The labourer now appears likely to obtain that hire of which he is worthy; and, therefore, this appears to me to be the time to press upon him the duty of using his best exertions for the rearing of his family in a manner that must give him the best security for happiness to himself, his wife and children, and to make him, in all respects, what his forefathers were. The people of England have been famed, in all ages, for their good living; for the abundance of their food, and goodness of their attire. The old sayings about English roast beef and plum-pudding, and about English hospitality, had not their foundation in *nothing*. And in spite of all refinements of sickly minds, it is *abundant living* amongst the people at large, which is the great test of good government, and the surest basis of national greatness and security.

9. If the labourer have his fair wages; if there be no false weights and measures, whether of money or of goods, by which he is defrauded; if the laws be equal in their effect upon all men, if he be called upon for no more than his due share of the expenses necessary to support the government and defend the

country, he has no reason to complain. If the largeness of his family demand extraordinary labour and care, these are due from him to it. He is the cause of the existence of that family; and, therefore, he is not, except in cases of accidental calamity, to throw upon others the burden of supporting it. Besides, " little children are as arrows in the hands of the " giant, and blessed is the man, that hath his quiver " full of them." That is to say, children, if they bring their cares, also bring their pleasures and solid advantages. They become, very soon, so many assistants and props to the parents, who, when old age comes on, are amply repaid for all the toils and all the cares that children have occasioned in their infancy. To be without sure and safe friends in the world makes life not worth having; and whom can we be so sure of as of our children? Brothers and sisters are a mutual support. We see them, in almost every case, grow up into prosperity, when they act the part that the impulses of nature prescribe. When cordially united, a father and sons, or a family of brothers and sisters, may, in almost any state of life, set what is called misfortune at defiance.

10. These considerations are much more than enough to sweeten the toils and cares of parents, and to make them regard every additional child as an additional blessing. But, that children may be a blessing and not a curse, care must be taken of their education. This word has, of late years, been so perverted, so corrupted, so abused, in its application, that I am almost afraid to use it here. Yet I must not suffer it to be usurped by cant and tyranny. I

must use it; but not without clearly saying what I mean.

11. Education means breeding up, bringing up or rearing up; and nothing more. This includes every thing with regard to the mind as well as the body of a child; but, of late years, it has been so used as to have no sense applied to it but that of book-learning, with which, nine times out of ten, it has nothing at all to do. It is, indeed, proper, and it is the duty of all parents to teach, or cause to be taught, their children as much as they can of books, after, and not before, all the measures are safely taken for enabling them to get their living by labour, or for providing them a living without labour, and that, too, out of the means obtained and secured by the parents out of their own income. The taste of the times is, unhappily, to give to children something of book-learning, with a view of placing them to live, in some way or other, upon the labour of other people. Very seldom, comparatively speaking, has this succeeded, even during the wasteful public expenditure of the last thirty years; and, in the times that are approaching, it cannot, I thank God, succeed at all. When the project has failed, what disappointment, mortification and, misery, to both parent and child! The latter is spoiled as a labourer; his book-learning has only made him conceited: into some course of desperation he falls; and the end is but too often not only wretched but ignominious.

12. Understand me clearly here, however; for it is the duty of parents to give, if they be able, book-

learning to their children, having first taken care to make them capable of earning their living by bodily labour. When that object has once been secured, the other may, if the ability remain, be attended to. But I am wholly against children wasting their time in the idleness of what is called education; and particularly in schools over which the parents have no control, and where nothing is taught but the rudiments of servility, pauperism, and slavery.

13. The education that I have in view is, therefore, of a very different kind. You should bear constantly in mind, that nine-tenths of us are, from the very nature and necessities of the world, born to gain our livelihood by the sweat of our brow. What reason have we, then, to presume, that our children are not to do the same? If they be, as now and then one will be, endued with extraordinary powers of mind, those powers may have an opportunity of developing themselves; and if they never have that opportunity, the harm is not very great to us or to them. Nor does it hence follow that the descendants of labourers are always to be labourers. The path upwards is steep and long, to be sure. Industry, care, skill, excellence, in the present parent, lay the foundation of a rise, under more favourable circumstances, for his children. The children of these take another rise; and, by-and-by, the descendants of the present labourer become gentlemen.

14. This is the natural progress. It is by attempting to reach the top at a single leap that so much misery is produced in the world; and the propensity to make such attempts has been cherished and

encouraged by the strange projects that we have witnessed of late years for making the labourers virtuous and happy by giving them what is called education. The education which I speak of consists in bringing children up to labour with steadiness, with care, and with skill; to show them how to do as many useful things as possible; to teach them to do them all in the best manner; to set them an example in industry, sobriety, cleanliness, and neatness; to make all these habitual to them, so that they never shall be liable to fall into the contrary, to let them always see a good living proceeding from labour, and thus to remove from them the temptation to get at the goods of others by violent or fraudulent means, and to keep far from their minds all the inducements to hypocrisy and deceit.

15. And, bear in mind, that if the state of the labourer has its advantages, when compared with other callings and conditions of life, it has also its advantages. It is free from the torments of ambition, and from a great part of the causes of ill-health, for which not all the riches in the world and all the circumstances of high rank are a compensation. The able and prudent labourer is always safe, at the least; and that is what few men are who are lifted above him. They have losses and crosses to fear, the very thought of which never enters his mind, if he act well his part towards himself, his family, and his neighbour.

16. But, the basis of good to him, is steady and skilful labour. To assist him in the pursuit of this labour, and in the turning of it to the best account,

are the principal objects of the present little work. I propose to treat of brewing Beer, making Bread, keeping Cows and Pigs, rearing Poultry, and of other matters; and to show that, while, from a very small piece of ground, a large part of the food of a considerable family may be raised, the very act of raising it will be the best possible foundation of education of the children of the labourer; that it will teach them a great number of useful things, add greatly to their value when they go forth from their father's home, make them start in life with all possible advantages, and give them the best chance of leading happy lives. And is it not much more rational for parents to be employed in teaching their children how to cultivate a garden, to feed and rear animals, to make bread, beer, bacon, butter, and cheese, and to be able to do these things for themselves, or for others, than to leave them to prowl about the lanes and commons, or to mope at the heels of some crafty, sleek-headed pretended saint, who while he extracts the last penny from their pockets, bids them be contented with their misery, and promises them, in exchange for their pence, everlasting glory in the world to come? It is upon the hungry and wretched that the fanatic works. The dejected and forlorn are his prey. As an ailing carcass engenders vermin, a pauperised community engenders teachers of fanaticism, the very foundation of whose doctrines is, that we are to care nothing about this world, and that all our labours and exertions are in vain.

17. The man who is doing well, who is in good

health, who has a blooming and dutiful and cheerful
and happy family about him, and who passes his day
of rest amongst them, is not to be made to believe
that he was born to be miserable, and that poverty,
the natural and just reward of laziness, is to secure
him a crown of glory. Far be it from me to recommend a disregard of even outward observances as
to matters of religion; but, can it be religion to
believe that God hath made us to be wretched and
dejected? Can it be religion to regard, as marks of
His grace, the poverty and misery that almost in-
variably attend our neglect to use the means of
obtaining a competence in worldly things? Can it
be religion to regard as blessings those things, those
very things, which God expressly numbers amongst
his curses? Poverty never finds a place amongst the
blessings promised by God. His blessings are of a
directly opposite description; flocks, herds, corn,
wine, and oil; a smiling land; a rejoicing people;
abundance for the body and gladness of the heart:
these are the blessings which God promises to the
industrious, the sober, the careful, and the upright.
Let no man, then, believe that, to be poor and
wretched is a mark of God's favour; and let no man
remain in that state if he, by any honest means, can
rescue himself from it.

18. Poverty leads to all sorts of evil consequences.
Want, horrid want, is the great parent of crime. To
have a dutiful family, the father's principle of rule
must be love, not fear. His sway must be gentle, or
he will have only an unwilling and short-lived
obedience. But it is given to but few men to be gentle

and good-humoured amidst the various torments
attendant on pinching poverty. A competence is,
therefore, the first thing to be thought of; it is the
foundation of all good in the labourer's dwelling;
without it little but misery can be expected. " Health,
peace, and competence," one of the wisest of men
regards as the only things needful to man; but the
two former are scarcely to be had without the latter.
Competence is the foundation of happiness and of
exertion. Beset with wants, having a mind continually
harassed with fears of starvation, who can act with
energy, who can calmly think? To provide a good
living, therefore, for himself and family, is the very
first duty of every man. " Two things," says AGUR,
" have I asked; deny me them not before I die:
" remove far from me vanity and lies; give me
" neither poverty nor riches; feed me with food
" convenient for me: lest I be full and deny thee;
" or lest I be poor and steal."

19. A good living, therefore, a competence, is
the first thing to be desired and to be sought after;
and, if this little work should have the effect of aiding
only a small portion of the labouring classes in
securing that competence, it will afford great grati-
fication to their friend,

WM. COBBETT.

Kensington, 19*th July* 1821.

BREWING BEER

20. BEFORE I proceed to give any directions about brewing, let me mention some of the inducements to do the thing. In former times, to set about to show to Englishmen that it was good for them to brew beer in their houses, would have been as impertinent as gravely to insist that they ought to endeavour not to lose their breath; for, in those times (only forty years ago), to have a house and not to brew was a rare thing indeed. Mr. ELLMAN, an old man and a large farmer, in Sussex, has recently given in evidence, before a Committee of the House of Commons, this fact; that, forty years ago, there was not a labourer in his parish that did not brew his own beer; and that now there is not one that does it, except by chance the malt be given him. The causes of this change have been the lowering of the wages of labour, compared with the price of provisions, by the means of the paper-money; the enormous tax upon the barley when made into malt; and the increased tax upon hops. These have quite changed the customs of the English people as to their drink. They still drink beer, but in general it is of the brewing of common brewers, and in public-houses, of which the common brewers have become the owners, and have thus, by the aid of paper-money, obtained a monopoly in the supplying of the great body of the people with one of those things which, to the hard-working man, is almost a necessary of life.

21. These things will be altered. They must be

altered. The nation must be sunk into nothingness, or a new system must be adopted: and the nation will not sink into nothingness. The malt now pays a tax of 4s. 6d. a bushel, and the barley costs only 3s. This brings the bushel of malt to 8s. including the maltster's charge for malting. If the tax were taken off the malt, malt would be sold, at the present price of barley, for about 3s. 3d. a bushel; because a bushel of barley makes more than a bushel of malt, and the tax, besides its amount, causes great expenses of various sorts to the maltster. The hops pay a tax of 2d. a pound; and a bushel of malt requires, in general, a pound of hops; if these two taxes were taken off, therefore, the consumption of barley and of hops would be exceedingly increased; for double the present quantity would be demanded, and the land is always ready to send it forth.

22. It appears impossible that the landlords should much longer submit to these intolerable burdens on their estates. In short, they must get off the malt tax or lose those estates. They must do a great deal more, indeed; but that they must do at any rate. The paper-money is fast losing its destructive power; and things are, with regard to the labourers, coming back to what they were forty years ago, and therefore we may prepare for the making of beer in our own houses, and take leave of the poisonous stuff served out to us by common brewers. We may begin immediately; for, even at present prices, home brewed beer is the cheapest drink that a family can use, except milk, and milk can be applicable only in certain cases.

23. The drink which has come to supply the place of beer has, in general, been tea. It is notorious that tea has no useful strength in it; that it contains nothing nutritious; that it, besides being good for nothing, has badness in it, because it is well known to produce want of sleep in many cases, and in all cases, to shake and weaken the nerves. It is, in fact, a weaker kind of laudanum, which enlivens for the moment and deadens afterwards. At any rate it communicates no strength to the body; it does not in any degree assist in affording what labour demands. It is, then, of no use. And now, as to its cost, compared with that of beer. I shall make my comparison applicable to a year, or three hundred and sixty-five days. I shall suppose the tea to be only five shillings the pound, the sugar only sevenpence, the milk only twopence a quart. The prices are at the very lowest. I shall suppose a tea-pot to cost a shilling, six cups and saucers two shillings and sixpence, and six pewter spoons eighteen-pence. How to estimate the firing I hardly know, but certainly there must be in the course of the year two hundred fires made that would not be made, were it not for tea drinking. Then comes the great article of all, the time employed in this tea-making affair. It is impossible to make a fire, boil water, make the tea, drink it, wash up the things, sweep up the fire-place, and put all to rights again in a less space of time, upon an average, than two hours. However, let us allow one hour; and here we have a woman occupied no less than three hundred and sixty-five hours in the year; or thirty whole days at twelve hours in the day; that

is to say, one month out of the twelve in the year, besides the waste of the man's time in hanging about waiting for the tea! Needs there any thing more to make us cease to wonder at seeing labourers' children with dirty linen and holes in the heels of their stockings? Observe, too, that the time thus spent is, one half of it, the best time of the day. It is the top of the morning, which, in every calling of life, contains an hour worth two or three hours of the afternoon. By the time that the clattering tea-tackle is out of the way, the morning is spoiled, its prime is gone, and any work that is to be done afterwards lags heavily along. If the mother have to go out to work, the tea affair must all first be over. She comes into the field, in summer time, when the sun has gone a third part of his course. She has the heat of the day to encounter, instead of having her work done and being ready to return home at an early hour. Yet early she must go too; for there is the fire again to be made, the clattering tea-tackle again to come forward; and even in the longest day she must have candle light, which never ought to be seen in a cottage (except in case of illness) from March to September.

24. Now, then, let us take the bare cost of the use of tea. I suppose a pound of tea to last twenty days, which is not nearly half an ounce every morning and evening. I allow for each mess half a pint of milk. And I allow three pounds of the red dirty sugar to each pound of tea. The account of expenditure would then stand very high; but to these must be added the amount of the tea-tackle, one set of

which will, upon an average, be demolished every
year. To these outgoings must be added the cost of
beer at the public-house; for some the man will have,
after all, and the woman too, unless they be upon
the point of actual starvation. Two pots a week is
as little as will serve in this way; and here is a dead
loss of ninepence a week, seeing that two pots of
beer, full as strong, and a great deal better, can be
brewed at home for threepence. The account of the
year's tea drinking will then stand thus:—

	£	s.	d.
18lb. of tea	4	10	0
54lb. of sugar	1	11	6
365 pints of milk	1	10	0
Tea-tackle	0	5	0
200 fires	0	16	8
30 days' work	0	15	0
Loss by going to public-house ..	1	19	0
	£11	7	2

25. I have here estimated every thing at its very
lowest. The entertainment which I have here provided is as poor, as mean, as miserable, as any thing
short of starvation can set forth; and yet the wretched
thing amounts to a good third part of a good and
able labourer's wages! For this money he and his
family may drink good and wholesome beer; in a
short time, out of the mere savings from this waste,
may drink it out of silver cups and tankards. In a
labourer's family, wholesome beer, that has a little

life in it, is all that is wanted in general. Little children, that do not work, should not have beer. Broth, porridge, or something in that way, is the thing for them. However, I shall suppose, in order to make my comparison as little complicated as possible, that he brews nothing but beer as strong as the generality of beer to be had at the public-house, and divested of the poisonous drugs which that beer but too often contains; and I shall further suppose that he uses in his family two quarts of this beer every day from the first of October to the last day of March inclusive; three quarts a day during the months of June and September; and five quarts a day during the months of July and August; and if this be not enough, it must be a family of drunkards. Here are 1,097 quarts, or 274 gallons. Now, a bushel of malt will make eighteen gallons of better beer than that which is sold at the public-houses. And this is precisely a gallon for the price of a quart. People should bear in mind, that the beer bought at the public-house is loaded with a beer tax, with the tax on the public-house keeper, in the shape of license, with all the taxes and expenses of the brewer, and with all taxes, rent, and other expenses of the publican, and with all the profits of both brewer and publican; so that when a man swallows a pot of beer at a public-house, he has all these expenses to help to defray, besides the mere tax on the malt and on the hops.

26. Well, then, to brew this ample supply of good beer for a labourer's family, these 274 gallons, requires fifteen bushels of malt and (for let us do the

thing well) fifteen pounds of hops. The malt is now eight shillings a bushel, and very good hops may be bought for less than a shilling a pound. The grains and yeast will amply pay for the labour and fuel employed in the brewing; seeing that there will be pigs to eat the grains, and bread to be baked with the yeast. The account will then stand thus :—

	£	s.	d.
15 bushels of malt	6	0	0
15 pounds of hops	0	15	0
Wear of utensils	0	10	0
	£7	5	0

27. Here, then, is the sum of four pounds two shillings and twopence saved every year. The utensils for brewing are, a brass kettle, a mashing tub, coolers (for which washing tubs may serve), a half hogshead, with one end taken out, for a tun tub, about four nine-gallon casks, and a couple of eighteen-gallon casks. This is an ample supply of utensils, each of which will last, with proper care, a good long lifetime or two, and the whole of which, even if purchased new from the shop, will only exceed by a few shillings, if they exceed at all, the amount of the saving, arising *the very first year*, from quitting the troublesome and pernicious practice of drinking tea. The saving of each succeeding year would, if you chose it, purchase a silver mug to hold half a pint at least. However, the saving would naturally be applied to purposes more conducive to the well-being and happiness of a family.

28. It is not, however, the mere saving to which I look. This is, indeed, a matter of great importance, whether we look at the amount itself, or at the ultimate consequences of a judicious application of it; for four pounds make a great hole in a man's wages for the year; and when we consider all the advantages that would arise to a family of children from having these four pounds, now so miserably wasted, laid out upon their backs, in the shape of a decent dress, it is impossible to look at this waste without feelings of sorrow, not wholly unmixed with those of a harsher description.

29. But I look upon the thing in a still more serious light. I view the tea drinking as a destroyer of health, an enfeebler of the frame, an engenderer of effeminacy and laziness, a debaucher of youth and a maker of misery for old age. In the fifteen bushels of malt there are 570 pounds weight of sweet; that is to say, of nutritious matter, unmixed with any thing injurious to health. In the 730 tea messes of the year there are 54 pounds of sweet in the sugar, and about 30 pounds of matter equal to sugar in the milk. Here are eighty-four pounds instead of five hundred and seventy, and even the good effect of these eighty-four pounds is more than over-balanced by the corrosive, gnawing, and poisonous powers of the tea.

30. It is impossible for any one to deny the truth of this statement. Put it to the test with a lean hog: give him the fifteen bushels of malt, and he will repay you in ten score of bacon or thereabouts. But give him the 730 tea messes, or rather begin to give them to him, and give him nothing else, and he is

dead with hunger, and bequeaths you his skeleton, at the end of about seven days. It is impossible to doubt in such a case. The tea drinking has done a great deal in bringing this nation into the state of misery in which it now is; and the tea drinking, which is carried on by " dribs " and " drabs ," by pence and farthings going out at a time; this miserable practice has been gradually introduced by the growing weight of the taxes on malt and on hops, and by the everlasting penury amongst the labourers, occasioned by the paper-money.

31. We see better prospects, however, and therefore let us now rouse ourselves, and shake from us the degrading curse, the effects of which have been much more extensive and infinitely more mischievous than men in general seem to imagine.

32. It must be evident to every one, that the practice of tea drinking must render the frame feeble and unfit to encounter hard labour or severe weather, while, as I have shown, it deducts from the means of replenishing the belly and covering the back. Hence succeeds a softness, an effeminacy, a seeking for the fire-side, a lurking in the bed, and, in short, all the characteristics of idleness, for which, in this case, real want of strength furnishes an apology. The tea drinking fills the public-house, makes the frequenting of it habitual, corrupts boys as soon as they are able to move from home, and does little less for the girls, to whom the gossip of the tea-table is no bad preparatory school for the brothel. At the very least, it teaches them idleness. The everlasting dawdling about with the slops of the tea-tackle gives

them a relish for nothing that requires strength and activity. When they go from home, they know how to do nothing that is useful. To brew, to bake, to make butter, to milk, to rear poultry; to do any earthly thing of use they are wholly unqualified. To shut poor young creatures up in manufactories is bad enough: but there, at any rate, they do something that is useful; whereas the girl that has been brought up merely to boil the tea-kettle, and to assist in the gossip inseparable from the practice, is a mere consumer of food, a pest to her employer, and a curse to her husband, if any man be so unfortunate as to fix his affections upon her.

33. But is it in the power of any man, any good labourer who has attained the age of fifty, to look back upon the last thirty years of his life, without cursing the day in which tea was introduced into England? Where is there such a man, who cannot trace to this cause a very considerable part of all the mortifications and sufferings of his life? When was he ever too late at his labour; when did he ever meet with a frown, with a turning off, and pauperism on that account, without being able to trace it to the tea-kettle? When reproached with lagging in the morning, the poor wretch tells you that he will make up for it by working during his breakfast time! I have heard this a hundred and a hundred times over. He was up time enough; but the tea-kettle kept him lolling and lounging at home; and now instead of sitting down to a breakfast upon bread, bacon, and beer, which is to carry him on to the hour of dinner, he has to force his limbs along under the sweat of

feebleness, and at dinner-time to swallow his dry bread, or slake his half-feverish thirst at the pump or the brook. To the wretched tea-kettle he has to return at night, with legs hardly sufficient to maintain him: and thus he makes his miserable progress towards that death which he finds ten or fifteen years sooner than he would have found it had he made his wife brew beer instead of making tea. If he now and then gladdens his heart with the drugs of the public-house, some quarrel, some accident, some illness, is the probable consequence; to the affray abroad succeeds an affray at home; the mischievous example reaches the children, corrupts them or scatters them, and misery for life is the consequence.

34. I shall now proceed to the details of brewing; but these, though they will not occupy a large space, must be put off to the second number. The custom of brewing at home has so long ceased, amongst labourers, and, in many cases, amongst tradesmen, that it was necessary for me fully to state my reasons for wishing to see the custom revived. I shall, in my next, clearly explain how the operation is performed; and it will be found to be so easy a thing, that I am not without hope, that many tradesmen, who now spend their evenings at a public-house, amidst tobacco smoke and empty noise, may be induced, by the finding of better drink at home, at a quarter part of the price, to perceive that home is by far the pleasantest place wherein to pass their hours of relaxation.

35. My work is intended chiefly for the benefit of

cottagers, who must, of course, have some land; for, I purpose to show, that a large part of the food of even a large family may be raised, without any diminution of the labourer's earnings abroad, from forty rod, or a quarter of an acre, of ground; but, at the same time, what I have to say will be applicable to larger establishments, in all the branches of domestic economy; and especially to that of providing a family with beer.

36. The kind of beer for a labourer's family, that is to say, the degree of strength, must depend on circumstances; on the numerousness of the family; on the season of the year, and various other. But, generally speaking, beer half the strength of that mentioned in Paragraph 25 will be quite strong enough; for that is, at least, one-third stronger than the farmhouse "small-beer," which, however, as long experience has proved, is best suited to the purpose. A judicious labourer would probably always have some ale in his house, and have small beer for the general drink. There is no reason why he should not keep Christmas as well as the farmer; and when he is mowing, reaping, or is at any other hard work, a quart, or three pints, of really good fat ale a-day is by no means too much. However, circumstances vary so much with different labourers, that, as to the sort of beer, and the number of brewings, and the times of brewing, no general rule can be laid down.

37. Before I proceed to explain the uses of the several brewing utensils, I must speak of the quality of the materials of which beer is made; that is to say,

the malt, hops, and water. Malt varies very much in quality, as, indeed, it must, with the quality of the barley. When good, it is full of flour, and in biting a grain asunder, you find it bite easily, and see the shell thin and filled up with flour. If it bite hard and steely, the malt is bad. There is pale malt and brown malt; but the difference in the two arises merely from the different degrees of heat employed in the drying. The main thing to attend to is, the quantity of flour. If the barley is bad, thin or steely, whether from unripeness or blight, or any other cause, it will not malt so well; that is to say, it will not send out its roots in due time; and a part of it will still be barley. Then, the world is wicked enough to think, and even to say, that there are maltsters who, when they send you a bushel of malt, put a little barley amongst it, the malt being taxed and the barley not! Let us hope that this is seldom the case; yet, when we do know that this terrible system of taxation induces the beer-selling gentry to supply their customers with stuff little better than poison, it is not very uncharitable to suppose it possible for some maltsters to yield to the temptations of the Devil so far as to play the trick above mentioned. To detect this trick, and to discover what portion of the barley is in an unmalted state, take a handful of the unground malt, and put it into a bowl of cold water. Mix it about with the water a little; that is, let every grain be just wet all over; and whatever part of them sink are not good. If you have your malt ground, there is not, as I know of, any means of detection. Therefore, if your brewing be considerable in amount, grind your own malt, the means of doing which is very easy, and neither expensive nor troublesome, as will appear, when I come to speak of flour. If the barley be well malted, there is still a variety in the quality of the malt; that is to say, a bushel of malt from fine, plump, heavy barley, will be better than the same quantity from thin and light barley. In this case, as in the case of wheat, the weight is the criterion of the quality. Only bear in mind, that as a bushel of wheat, weighing sixty-two pounds, is better worth six shillings, than a bushel weighing fifty-two pounds is worth four shillings, so a bushel of malt weighing forty-five pounds is better worth nine shillings, than a bushel weighing thirty-five is worth six shillings. In malt, therefore, as in every thing else, the word cheap is a deception, unless the quality be taken into view. But, bear in mind, that in the case of unmalted barley, mixed with the malt, the weight can be no rule; for barley is heavier than malt.

No. II

BREWING BEER

(Continued.)

38. As to using barley in the making of beer, I have given it a full and fair trial twice over, and I would recommend it to neither rich nor poor. The barley produces strength, though nothing like the

malt; but the beer is flat, even though you use half malt and half barley; and flat beer lies heavy on the stomach, and of course, besides the bad taste, is unwholesome. To pay 4s. 6d. tax upon every bushel of our own barley turned into malt, when the barley itself is not worth 3s. a bushel, is a horrid thing; but, as long as the owners of the land shall be so dastardly as to suffer themselves to be thus deprived of the use of their estates to favour the slave-drivers and plunderers of the East and West Indies, we must submit to the thing, incomprehensible to foreigners, and even to ourselves, as the submission may be.

39. With regard to hops, the quality is very various. At times when some sell for 5s. a pound, others sell for sixpence. Provided the purchaser understands the article, the quality is, of course, in proportion to the price. There are two things to be considered in hops; the power of preserving beer, and that of giving it a pleasant flavour. Hops may be strong, and yet not good. They should be bright, have no leaves or bits of branches amongst them. The hop is the husk, or seed-pod, of the hop-vine, as the cone is that of the fir-tree; and the seeds themselves are deposited, like those of the fir, round a little soft stalk, enveloped by the several folds of this pod or cone. If, in the gathering, leaves of the vine or bits of the branches are mixed with the hops, these not only help to make up the weight, but they give a bad taste to the beer; and, indeed, if they abound much, they spoil the beer. Great attention is therefore necessary in this respect. There are, too, numerous sorts of hops, varying in size, form, and

quality, quite as much as apples. However, when they are in a state to be used in brewing, the marks of goodness are, an absence of brown colour (for that indicates perished hops); a colour between green and yellow; a great quantity of the yellow farina; seeds not too large nor too hard; a clammy feel when rubbed between the fingers; and a lively, pleasant smell. As to the age of hops, they retain for twenty years, probably, their power of preserving beer; but not of giving it a pleasant flavour. I have used them at ten years old, and should have no fear of using them at twenty. They lose none of their bitterness; none of their power of preserving beer; but they lose the other quality; and therefore, in the making of fine ale, or beer, new hops are to be pre-ferred. As to the quantity of hops, it is clear, from what has been said, that that must, in some degree, depend upon their quality; but, supposing them to be good in quality, a pound of hops to a bushel of malt is about the quantity. A good deal, however, depends upon the length of time that the beer is intended to be kept, and upon the season of the year in which it is brewed. Beer intended to be kept a long while should have the full pound, also beer brewed in warmer weather, though for present use: half the quantity may do under an opposite state of circumstances.

40. The water should be soft by all means. That of brooks, or rivers, is best. That of a pond, fed by a rivulet, or spring, will do very well. Rainwater, if just fallen, may do; but stale rain-water, or stagnant pond-water, makes the beer flat and difficult to

keep; and hard water, from wells, is very bad: it does not get the sweetness out of the malt, nor the bitterness out of the hops, like soft water; and the wort of it does not ferment well, which is a certain proof of its unfitness for the purpose.

41. There are two descriptions of persons whom I am desirous to see brewing their own beer; namely, tradesmen, and labourers and journeymen. There must, therefore, be two distinct scales treated of. In the former editions of this work, I spoke of a machine for brewing, and stated the advantages of using it in a family of any considerable consumption of beer; but, while, from my desire to promote private brewing, I strongly recommended the machine, I stated, that, " if any of my readers " could point out any method by which we should " be more likely to restore the practice of private " brewing, and especially the cottage, I should be " greatly obliged to them to communicate it to me." Such communications have been made, and I am very happy to be able, in this new edition of my little work, to avail myself of them. There was, in the Patent Machine, always an objection on account of the expense; for, even the machine for one bushel of malt cost, at the reduced price, eight pounds; a sum far above the reach of the cottager, and even above that of a small tradesman. Its convenience, especially in towns, where room is so valuable, was an object of great importance; but there were disadvantages attending it which, until after some experience, I did not ascertain. It will be remembered that the method by the brewing machine

requires the malt to be put into the cold water, and for the water to make the malt swim, or, at least, to be in such proportion as to prevent the fire beneath from burning the malt. We found that our beer was flat, and that it did not keep. And this arose, I have every reason to believe, from this process. The malt should be put into hot water, and the water, at first, should be but just sufficient in quantity to stir the malt in, and separate it well. Nevertheless, when it is merely to make small beer; beer not wanted to keep; in such cases the brewing machine may be of use; and, as it will be seen by-and-by, a moveable boiler (which has nothing to do with the patent) may, in many cases, be of great convenience and utility.

42. The two scales of which I have spoken above are now to be spoken of; and, that I may explain my meaning the more clearly, I shall suppose, that, for the tradesman's family, it will be requisite to brew eighteen gallons of ale and thirty-six of small beer, to fill three casks of eighteen gallons each. It will be observed, of course, that, for larger quantities, larger utensils of all sorts will be wanted. I take this quantity as the one to give directions on. The utensils wanted here will be, FIRST, a copper that will contain forty gallons at least; for, though there be to be but thirty-six gallons of small beer, there must be space for the hops, and for the liquor that goes off in steam. SECOND, a mashing-tub, to contain sixty gallons; for the malt is to be in this along with the water. THIRD, an underbuck or shallow tub to go under the mash-tub, for the wort to run into when

drawn from the grains. FOURTH, a tun-tub, that will contain thirty gallons, to put the ale into to work, the mash-tub, as we shall see, serving as a tun-tub for the small beer. Besides these, a couple of coolers, shallow tubs, which may be the heads of wine butts, or some such things, about a foot deep; or if you have four it may be as well, in order to effect the cooling more quickly.

43. You begin by filling the copper with water, and next by making the water boil. You then put into the mashing-tub water sufficient to stir and separate the malt in. But now let me say more particularly what this mashing-tub is. It is, you know, to contain sixty gallons. It is to be a little broader at top than at bottom, and not quite so deep as it is wide across the bottom. In the middle of the bottom there is a hole about two inches over, to draw the wort off through. Into this hole goes a stick, a foot or two longer than the tub is high. This stick is to be about two inches through, and tapered for about eight inches upwards at the end that goes into the hole, which at last it fills up closely as a cork. Upon the hole, before any thing else be put into the tub, you lay a little bundle of fine birch (heath or straw may do), about half the bulk of a birchbroom, and well tied at both ends. This being laid over the hole (to keep back the grains as the wort goes out), you put the tapered end of the stick down through into the hole, and thus cork the hole up. You must then have something of weight sufficient to keep the birch steady at the bottom of the tub, with a hole through it to slip down the stick; otherwise when the stick is

raised it will be apt to raise the birch with it, and when you are stirring the mash you would move it from its place. The best thing for this purpose will be a leaden collar for the stick, with the hole in the collar plenty large enough, and it should weigh three or four pounds. The thing they use in some farm houses is the iron box of a wheel. Any thing will do that will slide down the stick and lie with weight enough on the birch to keep it from moving. Now, then, you are ready to begin brewing. I allow two bushels of malt for the brewing I have supposed. You must now put into the mashing-tub as much boiling water as will be sufficient to stir the malt in and separate it well. But here occur some of the nicest points of all: namely, the degree of heat that the water is to be at, before you put in the malt. This heat is one hundred and seventy degrees by the thermometer. If you have a thermometer, this is ascertained easily; but, without one, take this rule, by which so much good beer has been made in England for hundreds of years: when you can, by looking down into the tub, see your face clearly in the water, the water is become cool enough; and you must not put the malt in before. Now put in the malt and stir it well in the water. To perform this stirring, which is very necessary, you have a stick, somewhat bigger than a broom stick, with two or three smaller sticks, eight or ten inches long, put through the lower end of it at about three or four inches asunder, and sticking out on each side of the long stick. These small cross sticks serve to search the malt, and separate it well in the stirring

or mashing. Thus, then, the malt is in; and in this
state it should continue for about a quarter of an
hour. In the meanwhile you will have filled up your
copper, and made it boil; and now (at the end of the
quarter of an hour) you put in boiling water sufficient to give you your eighteen gallons of ale. But,
perhaps, you must have thirty gallons of water in
the whole; for the grains will retain at least ten gallons
of water; and it is better to have rather too much
wort than too little. When your proper quantity of
water is in, stir the malt again well. Cover the mashing-tub over with sacks, or something that will
answer the same purpose; and there let the mash
stand for two hours. When it has stood the two hours,
you draw off the wort. And now, mind, the mashing-
tub is placed on a couple of stools, or on something
that will enable you to put the underbuck under it,
so as to receive the wort as it comes out of the hole
before mentioned. When you have put the underbuck in its place, you let out the wort by pulling up
the stick that corks the hole. But, observe, this stick
(which goes six or eight inches through the hole)
must be raised by degrees, and the wort must be
let out slowly, in order to keep back the sediment.
So that it is necessary to have something to keep the
stick up at the point where you are to raise it, and
wish to fix it at for the time. To do this, the simplest,
cheapest and best thing in the world is a cleft stick.
Take a rod of ash, hazel, birch, or almost any wood,
let it be a foot or two longer than your mashing-tub
is wide over the top; split it, as if for making hoops;
tie it round with a string at each end; lay it across

your mashing-tub; pull it open in the middle and let the upper part of the wort stick through it, and when you raise that stick, by degrees as before directed, the cleft stick will hold it up at whatever height you please.

44. When you have drawn off the ale-wort, you proceed to put into the mashing-tub water for the small beer. But, I shall go on with my directions about the ale till I have got it into the cask and cellar; and shall then return to the small beer.

45. As you draw off the ale-wort into the under-buck, you must lade it out of that into the tun-tub, for which work, as well as for various other purposes in the brewing, you must have a bowl-dish, with a handle to it. The underbuck will not hold the whole of the wort. It is, as before described, a shallow tub to go under the mashing-tub to draw off the wort into. Out of this underbuck you must lade the ale-wort into the tun-tub; and there it must remain till your copper be emptied and ready to receive it.

46. The copper being empty, you put the wort into it, and put in after the wort, or before it, a pound and a half of good hops, well rubbed and separated as you put them in. You now make the copper boil, and keep it, with the lid off, at a good brisk boil, for a full hour, and if it be an hour and a half it is none the worse.

47. When the boiling is done, put out your fire, and put the liquor into the coolers. But it must be put into the coolers without the hops. Therefore, in order to get the hops out of the liquor, you must have a strainer. The best for your purpose is a small

clothes-basket, or any other wicker-basket. You set your coolers in the most convenient place. It may be in-doors or out of doors, as most convenient. You lay a couple of sticks across one of the coolers, and put the basket upon them. Put your liquor, hops and all, into the basket, which will keep back the hops. When you have got liquor enough in one cooler, you go to another with your sticks and basket, till you have got all your liquor out. If you find your liquor deeper in one cooler than the other, you can make an alteration in that respect, till you have the liquor so distributed as to cool equally fast in both, or all, the coolers.

48. The next stage of the liquor is the tun-tub, where it is set to work. Now, a very great point is, the degree of heat that the liquor is to be at when it is set to working. The proper heat is seventy degrees; so that a thermometer makes this matter sure. In the country they determine the degree of heat by merely putting a finger into the liquor. Seventy degrees is but just warm, a gentle luke-warmth. Nothing like heat. A little experience makes perfectness in such a matter. When at the proper heat, or nearly (for the liquor will cool a little in being removed), put it into the tun-tub. And now, before I speak of the act of setting the beer to work, I must describe this tun-tub which I first mentioned in Paragraph 42. It is to hold thirty gallons, as you have seen; and nothing is better than an old cask of that size, or somewhat larger, with the head taken out, or cut off. But, indeed, any tub of sufficient dimensions and of about the same depth proportioned to the

width as a cask or barrel, will do for the purpose.
Having put the liquor into the tun-tub, you put in
the yeast. About half a pint of good yeast is sufficient.
This should first be put into a thing of some sort
that will hold about a gallon of your liquor; the
thing should then be nearly filled with liquor, and
with a stick or spoon you should mix the yeast well
with the liquor in this bowl or other thing, and stir
in along with the yeast a handful of wheat or rye
flour. This mixture is then to be poured out clean
into the tun-tub, and the whole mass of the liquor
is then to be agitated well by lading up and pouring
down again with your bowl-dish, till the yeast be
well mixed with the liquor. Some people do the
thing in another manner. They mix up the yeast
and flour with some liquor (as just mentioned) taken
out of the coolers; and then they set the little vessel
that contains this mixture down on the bottom of
the tun-tub; and, leaving it there, put the liquor
out of the coolers into the tun-tub. Being placed at
the bottom, and having the liquor poured on it, the
mixture is, perhaps, more perfectly effected in this
way than in any other way. The flour may not be
necessary; but, as the country people use it, it is,
doubtless, of some use; for their hereditary experience
has not been for nothing. When your liquor is thus
properly put into the tun-tub and set a working, cover
over the top of the tub by laying across it a sack or
two, or something that will answer the purpose.

49. We now come to the last stage; the cask or
barrel. But I must first speak of the place for the
tun-tub to stand in. The place should be such as to

avoid too much warmth or cold. The air should, if possible, be at about 55 degrees. Any cool place in summer and any warmish place in winter. If the weather be very cold, some cloths or sacks should be put round the tun-tub while the beer is working. In about six or eight hours a frothy head will rise upon the liquor; and it will keep rising, more or less slowly, for about forty-eight hours. But, the length of time required for the working depends on various circumstances; so that no precise time can be fixed. The best way is, to take off the froth (which is indeed yeast) at the end of about twenty-four hours, with a common skimmer, and put it into a pan or vessel of some sort; then, in twelve hours' time, take it off again in the same way: and so on till the liquor has done working, and sends up no more yeast. Then it is beer; and when it is quite cold (for ale or strong beer) put it into the cask by means of a funnel. It must be cold before you do this, or it will be what the country people call foxed; that is to say, have a rank and disagreeable taste. Now, as to the cask, it must be sound and sweet. I thought, when writing the former edition of this work, that the bell-shaped were the best casks. I am now convinced that that was an error. The bell-shaped, by contracting the width of the top of the beer, as that top descends in consequence of the draft for use, certainly prevents the head (which always gathers on beer as soon as you begin to draw it off) from breaking and mixing in amongst the beer. This is an advantage in the bell-shape; but then the bell-shape, which places the widest end of the cask uppermost, exposes the cask to the admission of external air much more than the other shape. This danger approaches from the ends of the cask: and, in the bell-shape, you have the broadest end wholly exposed the moment you have drawn out the first gallon of beer, which is not the case with the casks of the common shape. Directions are given, in the case of the bell-casks, to put damp sand on the top to keep out the air. But, it is very difficult to make this effectual; and yet, if you do not keep out the air, your beer will be flat; and when flat, is really is good for nothing but the pigs. It is very difficult to fill the bell-cask, which you will easily see if you consider its shape. It must be placed on the level with the greatest possible truth, or there will be a space left; and to place it with such truth is, perhaps, as difficult a thing as a mason or bricklayer ever had to perform. And yet, if this be not done, there will be an empty space in the cask, though it may, at the same time, run over. With the common casks, there are none of these difficulties. A common eye will see when it is well placed; and, at any rate, any little vacant space that may be left is not at an end of the cask, and will, without great carelessness, be so small as to be of no consequence. We now come to the act of putting in the beer. The cask should be placed on a stand with legs about a foot long. The cask, being round, must have a little wedge, or block, on each side to keep it steady. Bricks do very well. Bring your beer down into the cellar in buckets, and pour it in through the funnel, until the cask be full. The cask should lean a little on one side when you fill it; because the beer will work again here, and send more yeast out of the bung-hole; and, if the cask were not a little on one side, the yeast would flow over both sides of the cask, and would not descend in one stream into a pan, put underneath to receive it. Here the bell-cask is extremely inconvenient; for the yeast works up all over the head, and cannot run off, and makes a very nasty affair. This alone, to say nothing of the other disadvantages, would decide the question against the bell-casks. Something will go off in this working, which may continue for two or three days. When you put the beer in the cask, you should have a gallon or two left, to keep filling up with, as the working produces emptiness. At last, when the working is completely over, right the cask. That is to say, block it up to its level. Put in a handful of fresh hops. Fill the cask quite full. Put in the bung, with a bit of coarse linen stuff round it; hammer it down tight; and, if you like, fill a coarse bag with sand, and lay it, well pressed down, over the bung.

50. As to the length of time that you are to keep the beer before you begin to use it, that must, in some measure, depend on taste. Such beer as this ale, will keep almost any length of time. As to the mode of tapping, that is as easy almost as drinking. When the cask is empty, great care must be taken to cork it tightly up, so that no air get in; for, if it do, the cask is moulded, and when once moulded, it is spoiled for ever. It is never again fit to be used about beer. Before the cask be used again, the grounds must be poured out, and the cask cleaned

by several times scalding; by putting in stones (or a chain), and rolling and shaking about till it be quite clean. Here again the round casks have the decided advantage; it being almost impossible to make the bell-casks thoroughly clean, without taking the head out, which is both troublesome and expensive; as it cannot be well done by any one but a cooper, who is not always at hand, and who, when he is, must be paid.

51. I have now done with the ale, and it remains for me to speak of the small beer. In Paragraph 47 (which now see) I left you drawing off the ale-wort, and with your copper full of boiling water. Thirty-six gallons of that boiling water are, as soon as you have got your ale-wort out, and have put down your mash-tub stick to close up the hole at the bottom; as soon as you have done this, 36 gallons of the boiling water are to go into the mashing-tub; the grains are to be well stirred up, as before; the mashing-tub is to be covered over again as mentioned in Paragraph 43; and the mash is to stand in that state for an hour; and not two hours, as for the ale-wort.

52. When the small beer mash has stood its hour, draw it off as in Paragraph 47, and put it into the tun-tub as you did the ale-wort.

53. By this time your copper will be empty again, by putting your ale liquor to cool, as mentioned in Paragraph 47. And now you put the small beer wort into the copper, with the hops that you used before, and with half a pound of fresh hops added to them, and this liquor you boil briskly for an hour.

54. By this time you will have taken the grains

and the sediment clean out of the mashing-tub, and taken out the bunch of birch twigs, and made all clean. Now put in the birch twigs again, and put down your stick as before. Lay your two or three sticks across the mashing-tub, put your basket on them, and take your liquor from the copper (putting the fire out first) and pour it into the mashing-tub through the basket. Take the basket away, throw the hops to the dunghill, and leave the small beer liquid to cool in the mashing-tub.

55. Here it is to remain to be set to working as mentioned for the ale in Paragraph 48; only in this case, you will want more yeast in proportion; and should have for your 36 gallons of small beer, three half pints of good yeast.

56. Proceed, as to all the rest of the business, as with the ale; only, in the case of the small beer, it should be put into the cask, not quite cold but a little warm; or else it will not work at all in the barrel, which it ought to do. It will not work so strongly or so long as the ale; and may be put in the barrel much sooner; in general the next day after it is brewed.

57. All the utensils should be well cleaned and put away as soon as they are done with; the little things as well as the great things; for it is loss of time to make new ones. And, now, let us see the expense of these utensils. The copper, new, £5; the mashing-tub, new, 30s.; the tun-tub, not new, 5s.; the underbuck and three coolers, not new, 20s. The whole cost is £7 10s., which is ten shillings less than the one bushel machine. I am now in a farm house, where the same set of utensils has been used

for forty years; and the owner tells me, that, with
the same use, they may last for forty years longer.
The machine will not, I think, last four years, if in
anything like regular use. It is of sheet-iron, tinned
on the inside, and this tin rusts exceedingly, and is
not to be kept clean without such rubbing as must
soon take off the tin. The great advantage of the
machine is, that it can be removed. You can brew
without a brewhouse. You can set the boiler up
against any fireplace, or any window. You can brew
under a cart-shed, and even out of doors. But all
this may be done with these utensils, if your copper
be moveable. Make the boiler of copper, and not of
sheet iron, and fix it on a stand with a fire-place and
stove pipe ; and then you have the whole to brew
out of doors with as well as indoors, which is a very
great convenience.

58. Now, with regard to the other scale of brewing, little need be said; because all the principles
being the same, the utensils only are to be proportioned to the quantity. If only one sort of beer be
to be brewed at a time, all the difference is, that in
order to extract the whole of the goodness of the
malt, the mashing ought to be at twice. The two
worts are then put together, and then you boil them
together with the hops.

59. A Correspondent at Morpeth says, the whole
of the utensils used by him are a twenty-gallon pot,
a mashing-tub, that also answers for a tun-tub, and
a shallow tub for a cooler; and that these are plenty
for a person who is anything of a contriver. This is
very true; and these things will cost no more, per-

haps, than forty shillings. A nine-gallon cask of beer can be brewed very well with such utensils. Indeed, it is what used to be done by almost every labouring man in the kingdom, until the high price of malt and comparatively low price of wages rendered the people too poor and miserable to be able to brew at all. A Correspondent at Bristol has obligingly sent me the model of utensils for brewing on a small scale; but as they consist chiefly of brittle ware, I am of opinion that they would not so well answer the purpose.

60. Indeed, as to the country labourers, all they want is the ability to get the malt. Mr. ELLMAN, in his evidence before the Agricultural Committee, said, that, when he began farming, forty-five years ago, there was not a labourer's family in the parish that did not brew their own beer and enjoy it by their own fire-sides; and that now not one single family did it, from want of ability to get the malt. It is the tax that prevents their getting the malt, for the barley is cheap enough. The tax causes a monopoly in the hands of the maltsters, who, when the tax is two and sixpence, make the malt cost 7s. 6d., though the barley cost but 2s. 6d.; and though the malt, tax and all, ought to cost him about 5s. 6d. If the tax were taken off, this pernicious monopoly would be destroyed.

61. The reader will easily see, that in proportion to the quantity wanted to be brewed must be the size of the utensils; but I may observe here, that the above utensils are sufficient for three, or even four, bushels of malt, if stronger beer be wanted.

62. When it is necessary, in case of falling short in the quantity wanted to fill up the ale cask, some may be taken from the small beer. But, upon the whole brewing, there ought to be no falling short; because, if the casks be not filled up, the beer will not be good, and certainly will not keep. Great care should be taken as to the cleansing of the casks. They should be made perfectly sweet, or it is impossible to have good beer. •

63. The cellar, for beer to keep any length of time, should be cool. Under a hill is the best place for a cellar; but, at any rate, a cellar of good depth, and dry. At certain times of the year, beer that is kept long will ferment. The vent pegs must, in such cases, be loosened a little, and afterwards fastened.

64. Small beer may be tapped almost directly. It is a sort of a joke that it should see a Sunday, but that it may do before it be two days old. In short, any beer is better than water; but it should have some strength and some weeks of age, at any rate.

65. I cannot conclude this Essay without expressing my pleasure, that a law has been recently passed to authorise the general retail of beer. This really seems necessary to prevent the King's subjects from being poisoned. The brewers and porter quacks have carried their tricks to such an extent that there is no safety for those who drink brewers' beer.

66. The best and most effectual thing is, however, for people to brew their own beer, to enable them and induce them to do which I have done all that lies in my power. A longer treatise on the subject would have been of no use. These few plain direc-

tions will suffice for those who have a disposition to do the thing, and those who have not would remain unmoved by anything that I could say.

67. There seems to be a great number of things to do in brewing, but the greater part of them require only about a minute each. A brewing, such as I have given the detail of above, may be completed in a day; but, by the word day I mean to include the morning, beginning at four o'clock.

68. The putting of the beer into barrel is not more than an hour's work for a servant woman, or a tradesman's or farmer's wife. There is no heavy work, no work too heavy for a woman in any part of the business, otherwise I would not recommend it to be performed by the women, who, though so amiable in themselves, are never quite so amiable as when they are useful; and as to beauty, though men may fall in love with girls at play, there is nothing to make them stand to their love like seeing them at work. In conclusion of these remarks on beer brewing, I once more express my most anxious desire to see abolished for ever the accursed tax on malt, which, I verily believe, has done more harm to the people of England than was ever done to any people by plague, pestilence, famine, and civil war.

69. In Paragraph 76, in Paragraph 108, and perhaps in another place or two (of the last edition), I spoke of the machine for brewing. The work being stereotyped, it would have been troublesome to alter those paragraphs; but of course the public in reading them, will bear in mind what has been now said relative to the machine. The inventor of that machine

deserves great praise for his efforts to promote private brewing; and, as I said before, in certain confined situations, and where the beer is to be merely small beer, and for immediate use, and where time and room are of such importance as to make the cost of the machine comparatively of trifling consideration, the machine may possibly be found to be an useful utensil.

70. Having stated the inducements to the brewing of beer, and given the plainest directions that I was able to give for the doing of the thing, I shall next proceed to the subject of bread. But this subject is too large and of too much moment to be treated with brevity, and must therefore be put off till my next Number. I cannot, in the meanwhile, dismiss the subject of brewing beer without once more adverting to its many advantages, as set forth in the foregoing Number of this work.

71. The following instructions for the making of porter will clearly show what sort of stuff is sold at public-houses in London; and we may pretty fairly suppose that the public-house beer in the country is not superior to it in quality. " A quarter of malt, " with these ingredients, will make five barrels of " good porter. Take one quarter of high-coloured " malt, eight pounds of hops, nine pounds of treacle, " eight pounds of colour, eight pounds of sliced " liquorice root, two drachms of salt of tartar, two " ounces of Spanish-liquorice, and half an ounce of " capsicum." The author says, that he merely gives the ingredients as used by many persons.

72. This extract is taken from a book on brewing,

recently published in London. What a curious com-
position! What a mess of drugs! But, if the brewers
openly avow this, what have we to expect from the
secret practices of them and the retailers of the
article! When we know, that beer-doctor and
brewers'-druggist are professions, practised as
openly as those of bug-man and rat-killer, are we
simple enough to suppose that the above-named
are the only drugs that people swallow in these
potions, which they call pots of beer? Indeed we
know the contrary, for scarcely a week passes without witnessing the detection of some greedy wretch
who has used, in making or in doctoring his beer,
drugs forbidden by the law. And it is not many weeks
since one of these were convicted, in the Court of
Excise, for using a potent and dangerous drug, by
the means of which, and a suitable quantity of water,
he made two butts of beer into three. Upon this
occasion, it appeared that no less than ninety of
these worthies were in the habit of pursuing the
same practices. The drugs are not unpleasant to
the taste: they sting the palate: they give a present
relish: they communicate a momentary exhilara-
tion: but, they give no force to the body, which, on
the contrary, they enfeeble, and, in many instances,
with time, destroy; producing diseases from which
the drinker would otherwise have been free to the
end of his days.

73. But, look again at the receipt for making
porter. Here are eight bushels of malt for 180 gallons
of beer; that is to say, twenty-five gallons from the
bushel. Now the malt is eight shillings a bushel,

and eight pounds of the very best hops will cost but
a shilling a pound. The malt and hops, then, for the
180 gallons, cost but seventy-two shillings; that is
to say, only a little more than fourpence three farthings a gallon, for stuff which is now retailed for
sixteenpence a gallon! If this be not an abomination,
I should be glad to know what is. Even if the treacle,
colour, and the drugs, be included, the cost is not
fivepence a gallon; and yet not content with this
enormous extortion, there are wretches who resort
to the use of other and pernicious drugs, in order to
increase their gains!

74. To provide against this dreadful evil there is,
and there can be no law; for it is created by the law.
The law it is that imposes the enormous tax on the
malt and hops; the law it is that imposes the licence
tax, and places the power of granting the licence at
the discretion of persons appointed by the government; the law it is that checks, in this way, the private
brewing, and that prevents free and fair competition
in the selling of beer, and as long as the law does
these, it will in vain endeavour to prevent the people
from being destroyed by slow poison.

75. Innumerable are the benefits that would arise
from a repeal of the taxes on malt and on hops. Tip-
pling houses might then be shut up with justice and
propriety. The labourer, the artisan, the tradesman,
the landlord, all would instantly feel the benefit.
But the landlord more, perhaps, in this case, than
any other member of the community. The four or
five pounds a year which the day labourer now
drizzles away in tea-messes, he would divide with

the farmer, if he had untaxed beer. His wages would fall, and fall to his advantage too. The fall of wages would be not less than £40 upon a hundred acres. Thus £40 would go, in the end, a fourth perhaps to the farmer, and three-fourths to the landlord. This is a kind of work to reduce poor-rates, and to restore husbandry to prosperity. Undertaken this work must be, and performed too; but whether we shall see this until the estates have passed away from the present race of landlords, is a question which must be referred to time.

76. Surely we may hope, that, when the American farmers shall see this little Essay, they will begin seriously to think of leaving off the use of the liver-burning and palsy-producing spirits. Their climate, indeed, is something: extremely hot in one part of the year, and extremely cold in the other part of it. Nevertheless, they may have, and do have, very good beer if they will. Negligence is the greatest impediment in their way. I like the Americans very much; and that, if there were no other, would be a reason for my not hiding their faults.

No. III

MAKING BREAD

77. LITTLE time need be spent in dwelling on the necessity of this article to all families; though, on account of the modern custom of using potatoes to supply the place of bread, it seems necessary to say a few words here on the subject, which in another

work I have so amply, and I think so triumphantly discussed. I am the more disposed to revive the subject for a moment in this place, from having read in the evidence recently given before the Agricultural Committee, that many labourers, especially in the West of England, use potatoes instead of bread to a very great extent. And I find from the same evidence, that it is the custom to allot the labourers " a potato ground " in part payment of their wages! This has a tendency to bring English labourers down to the state of the Irish, whose mode of living, as to food, is but one remove from that of the pig, and of the ill-fed pig too.

78. I was, in reading the above-mentioned evidence, glad to find, that Mr. EDWARD WAKEFIELD, the best-informed and most candid of all the witnesses, gave it as his opinion, that the increase which had taken place in the cultivation of potatoes was " injurious to the country; " an opinion which must, I think, be adopted by every one who takes the trouble to reflect a little upon the subject. For leaving out of the question the slovenly and beastly habits engendered amongst the labouring classes by constantly lifting their principal food at once out of the earth to their mouths, by eating without the necessity of any implements other than the hands and teeth, and by dispensing with everything requiring skill in the preparation of the food, and requiring cleanliness in its consumption or preservation; leaving these out of the question, though they are all matters of great moment when we consider their effects in the rearing of a family, we shall find, that, in mere

quantity of food, that is to say, of nourishment, bread is the preferable diet.

79. An acre of land that will produce 300 bushels of potatoes will produce 32 bushels of wheat. I state this as an average fact, and am not at all afraid of being contradicted by any one well acquainted with husbandry. The potatoes are supposed to be of a good sort, as it is called, and the wheat may be supposed to weigh 60lb., a bushel. It is a fact clearly established, that, after the water, the stringy substance, and the earth, are taken from the potato, there remains only one tenth of the rough raw weight of nutritious matter, or matter which is deemed equally nutritious with bread, and as the raw potatoes weigh 56lb. a bushel, the acre will yield 1,830lb. of nutritious matter. Now mind, a bushel of wheat, weighing 60lb., will make of household bread (that is to say, taking out only the bran) 65lb. Thus, the acre yields 2,080lb. of bread. As to the expenses, the seed and act of planting are about equal in the two cases. But, while the potatoes must have cultivation during their growth, the wheat needs none; and while the wheat straw is worth from three to five pounds an acre, the haulm of the potatoes is not worth one single truss of that straw. Then, as to the expense of gathering, housing, and keeping the potato crop, it is enormous, besides the risk of loss by frost, which may be safely taken, on an average, at a tenth of the crop. Then comes the expense of cooking. The thirty-two bushels of wheat, supposing a bushel to be baked at a time (which would be the case in a large family), would demand thirty-two

heatings of the oven. Suppose a bushel of potatoes to be cooked every day in order to supply the place of this bread, then we have nine hundred boilings of the pot, unless cold potatoes be eaten at some of the meals; and in that case the diet must be cheering indeed! Think of the labour, think of the time, think of all the peelings and scrapings and washings and messings attending these nine hundred boilings of the pot! For it must be a considerable time before English people can be brought to eat potatoes in the Irish style; that is to say, scratch them out of the earth with their paws, toss them into the pot without washing, and when boiled turn them out upon a dirty board, and then sit round that board, peel the skin and dirt from one at a time and eat the inside. Mr. Curwen was delighted with " Irish hospitality," because the people there receive no parish relief; upon which I can only say, that I wish him the exclusive benefit of such hospitality.

80. I have here spoken of a large quantity of each of the sorts of food. I will now come to a comparative view, more immediately applicable to a labourer's family. When wheat is ten shillings the bushel, potatoes, bought at best hand (I am speaking of the country generally), are about two shillings a bushel. Last Spring the average price of wheat might be six and sixpence, and the average price of potatoes (in small quantities) was about eighteen-pence; though, by the waggon-load, I saw potatoes bought at a shilling a bushel, to give to sheep; then observe, these were of the coarsest kind, and the farmer had to fetch them at a considerable expense. I think,

therefore, that I give the advantage to the potatoes when I say that they sell, upon an average, for full a fifth part as much as the wheat sells for per bushel, while they contain four pounds less weight than the bushel of wheat; while they yield only five pounds and a half of nutritious matter equal to bread; and while the bushel of wheat will yield sixty-five pounds of bread, besides the ten pounds of bran. Hence it is clear that instead of the saving, which is everlastingly dinned in our ears, from the use of potatoes, there is a waste of more than one half; seeing that, when wheat is ten shillings the bushel you can have sixty-five pounds of bread for the ten shillings, and can have out of potatoes only five pounds and a half of nutritious matter equal to bread for two shillings! This being the case, I trust that we shall soon hear no more of those savings which the labourer makes by the use of potatoes; I hope we shall, in the words of Doctor DRENNAN, " leave Ireland to her lazy root," if she choose still to adhere to it. It is the root also of slovenliness, filth, misery, and slavery; its cultivation has increased. in England with the increase of the paupers: both, I thank God, are upon the decline! Englishmen seem to be on the return to beer and bread, from water and potatoes; and therefore I shall now proceed to offer some observations to the Cottager, calculated to induce him to bake his own bread.

81. As I have before stated, sixty pounds of wheat, that is to say, where the Winchester bushel weighs sixty pounds, will make sixty-five pounds of bread, besides the leaving of about ten pounds of bran.

This is household bread, made of flour from which the bran only is taken. If you make fine flour, you take out pollard, as they call it, as well as bran, and then you have a smaller quantity of bread and a greater quantity of offal; but even of this finer bread, bread equal in fineness to the baker's bread, you get from fifty-eight to fifty-nine pounds out of the bushel of wheat. Now then, let us see how many quartern loaves you get out of the bushel of wheat, supposing it to be fine flour, in the first place. You get thirteen quartern loaves and a half: these cost you, at the present average price of wheat (seven and sixpence a bushel), in the first place 7s. 6d.; then 3d. for yeast; then not more than 3d. for grinding, because you have about thirteen pounds of offal, which is worth more than a $\frac{1}{2}$d. a pound, while the grinding is 9d. a bushel. Thus then the bushel of bread of fifty-nine pounds cost you eight shillings, and it yields you the weight of thirteen and a half quartern loaves: these quartern loaves now (Dec., 1821) sell at Kensington, at the baker's shop, at 1s. $\frac{1}{2}$d.; that is to say, the thirteen quartern loaves and a half cost 14s. 7$\frac{1}{2}$d. I omitted to mention the salt, which would cost you 4d. more. So that here is 6s. $\frac{1}{2}$d. saved upon the baking of a bushel of bread. The baker's quartern loaf is indeed cheaper in the country than at Kensington, by probably a penny in the loaf; which would still, however, leave a saving of 5s. upon the bushel of bread. But, besides this, pray think a little of the materials of which the baker's loaf is composed. The alum, the ground potatoes, and other materials; it being a notorious fact, that the bakers, in London at least, have mills wherein to grind their potatoes, so large is the scale upon which they use that material. It is probable that, out of a bushel of wheat, they make between sixty and seventy pounds of bread, though they have no more flour, and of course no more nutritious matter, than you have in your fifty-nine pounds of bread. But at the least, supposing their bread to be as good as yours in quality, you have, allowing a shilling for the heating of the oven, a clear 4s. saved upon every bushel of bread. If you consume half a bushel a week, that is to say, about a quartern loaf a day, this is a saving of £5 4s. a year, or full a sixth part, if not a fifth part, of the earnings of a labourer in husbandry.

82. How wasteful then, and indeed how shameful, for a labourer's wife to go to the baker's shop; and how negligent, how criminally careless of the welfare of his family, must the labourer be, who permits so scandalous a use of the proceeds of his labour! But I have hitherto taken a view of the matter the least possibly advantageous to the home-baked bread. For, ninety-nine times out of a hundred, the fuel for heating the oven costs very little. The hedgers, the copsers, the woodmen of all descriptions, have fuel for little or nothing. At any rate, to heat the oven cannot, upon an average, take the country through, cost the labourer more than 6d. a bushel. Then, again, fine flour need not ever be used, and ought not to be used. This adds six pounds of bread to the bushel, or nearly another quartern loaf and a half, making nearly fifteen quartern loaves out of the bushel of wheat. The finest flour is by

no means the most wholesome; and, at any rate, there is more nutritious matter in a pound of household bread than in a pound of baker's bread. Besides this, rye, and even barley, especially when mixed with wheat, make very good bread. Few people upon the face of the earth live better than the Long Islanders. Yet nine families out of ten seldom eat wheaten-bread. Rye is the flour that they principally make use of. Now, rye is seldom more than two-thirds the price of wheat, and barley is seldom more than half the price of wheat. Half rye and half wheat, taking out a little more of the offal, make very good bread. Half wheat, a quarter rye and a quarter barley, nay, one-third of each, make bread that I could be very well content to live upon all my lifetime; and, even barley alone, if the barley be good, and none but the finest flour taken out of it, has in it, measure for measure, ten times the nutrition of potatoes. Indeed the fact is well known, that our forefathers used barley bread to a very great extent. Its only fault, with those who dislike it, is its sweetness, a fault which we certainly have not to find with the baker's loaf, which has in it little more of the sweetness of grain than is to be found in the offal which comes from the sawings of deal boards. The nutritious nature of barley is amply proved by the effect, and very rapid effect, of its meal, in the fatting of hogs and of poultry of all descriptions. They will fatten quicker upon meal of barley than upon any other thing. The flesh, too, is sweeter than that proceeding from any other food, with the exception of that which proceeds from buckwheat, a grain little used in England. That proceeding from Indian Corn is, indeed, still sweeter and finer; but this is wholly out of the question with us.

83. I am, by-and-bye, to speak of the cow to be kept by the labourer in husbandry. Then there will be milk to wet the bread with, an exceedingly great improvement in its taste as well as in its quality! This, of all the ways of using skim milk, is the most advantageous; and this great advantage must be wholly thrown away, if the bread of the family be bought at the shop. With milk, bread with very little wheat in it may be made far better than baker's bread; and, leaving the milk out of the question, taking a third of each sort of grain, you would get bread weighing as much as fourteen quartern loaves, for about 5s. 9d., at present prices of grain; that is to say, you would get it for about 5d. the quartern loaf, all expenses included; thus you have nine pounds and ten ounces of bread a day for about 5s. 9d. a week. Here is enough for a very large family. Very few labourers' families can want so much as this, unless indeed there be several persons in it capable of earning something by their daily labour. Here is cut and come again. Here is bread always for the table. Bread to carry a field; always a hunch of bread ready to put into the hand of a hungry child. We hear a great deal about " children crying for bread," and objects of compassion they and their parents are, when the latter have not the means of obtaining a sufficiency of bread. But I should be glad to be informed, how is it possible for a labouring man, who earns, upon an average, 10s. a week,

who has not more than four children (and if he have
more, some ought to be doing something); who has
a garden of a quarter of an acre of land (for that
makes part of my plan); who has a wife as industrious as she ought to be; who does not waste his
earnings at the ale house or the tea shop: I should
be glad to know how such a man, while wheat shall
be at the price of about 6s. a bushel, can possibly
have children crying for bread!

84. Cry, indeed, they must, if he will persist in
giving 13s. for a bushel of bread instead of 5s. 9d.
Such a man is not to say that the bread which I have
described is not good enough. It was good enough
for his forefathers, who were too proud to be paupers,
that is to say, abject and willing slaves. " Hogs eat
barley." And hogs will eat wheat, too, when they
can get it. Convicts in condemned cells eat wheaten
bread; but we think it no degradation to eat wheaten
bread, too. I am for depriving the labourer of none
of his rights; I would have him oppressed in no
manner or shape; I would have him bold and free;
but to have him such, he must have bread in his
house, sufficient for all his family, and whether that
bread be fine or coarse must depend upon the different circumstances which present themselves in the
cases of different individuals.

85. The married man has no right to expect the
same plenty of food and of raiment that the single
man has. The time before marriage is the time to
lay by, or if the party chose, to indulge himself in
the absence of labour. To marry is a voluntary act,
and it is attended in the result with great pleasures

and advantages. If, therefore, the laws be fair and equal; if the state of things be such that a labouring man can, with the usual ability of labourers, and with constant industry, care, and sobriety; with decency of deportment towards all his neighbours, cheerful obedience to his employer, and a due subordination to the laws; if the state of things be such, that such a man's earnings be sufficient to maintain himself and family with food, raiment, and lodging needful for them; such a man has no reason to complain; and no labouring man has reason to complain if the numerousness of his family should call upon him for extraordinary exertion, or for frugality uncommonly rigid. The man with a large family has, if it be not in a great measure his own fault, a greater number of pleasures and of blessings than other men. If he be wise, and just as well as wise, he will see that it is reasonable for him to expect less delicate fare than his neighbours, who have a less number of children, or no children at all. He will see the justice as well as the necessity of his resorting to the use of coarser bread, and thus endeavour to make up that, or at least a part of that, which he loses in comparison with his neighbours. The quality of the bread ought, in every case, to be proportioned to the number of the family and the means of the head of that family. Here is no injury to health proposed; but, on the contrary, the best security for its preservation. Without bread, all is misery. The Scripture truly calls it the staff of life; and it may be called, too, the pledge of peace and happiness in the labourer's dwelling.

86. As to the act of making bread, it would be shocking indeed if that had to be taught by the means of books. Every woman, high or low, ought to know how to make bread. If she do not, she is unworthy of trust and confidence: and, indeed, a mere burden upon the community. Yet it is but too true, that many women, even amongst those who have to get their living by their labour, know nothing of the making of bread, and seem to understand little more about it than the part which belongs to its consumption. A Frenchman, a Mr. CUSAR, who had been born in the West Indies, told me, that till he came to Long Island he never knew how the flour came; that he was surprised when he learnt that it was squeezed out of little grains that grew at the tops of straw; for that he had always had an idea that it was got out of some large substances, like the yams that grow in tropical climates. He was a very sincere and good man, and I am sure he told me truth. And this may be the more readily believed when we see so many women in England who seem to know no more of the constituent parts of a loaf than they know of those of the moon. Servant women in abundance appear to think that loaves are made by the baker, as knights are made by the king; things of their pure creation, a creation too in which no one else can participate. Now is not this an enormous evil? And whence does it come? Servant women are the children of the labouring classes; and they would all know how to make bread, and know well how to make it too, if they had been fed on bread of their mothers' and their own making.

87. How serious a matter, then, is this, even in this point of view! A servant that cannot make bread is not entitled to the same wages as one that can. If she can neither bake nor brew; if she be ignorant of the nature of flour, yeast, malt and hops, what is she good for? If she understand these matters well, if she be able to supply her employer with bread and with beer, she is really valuable; she is entitled to good wages, and to consideration and respect into the bargain; but if she be wholly deficient in these particulars, and can merely dawdle about with a bucket and a broom, she can be of very little consequence; to lose her is merely to lose a consumer of food, and she can expect very little indeed in the way of desire to make her life easy and pleasant. Why should any one have such a desire? She is not a child of the family. She is not a relation. Any one as well as she can take in a loaf from the baker, or a barrel of beer from the brewer. She has nothing whereby to bind her employer to her. To sweep a room anything is capable of that has got two hands. In short, she has no useful skill, no useful ability; she is an ordinary drudge, and she is treated accordingly.

88. But, if such be her state in the house of an employer, what is her state in the house of a husband? The lover is blind, but the husband has eyes to see with. He soon discovers that there is something wanted besides dimples and cherry cheeks; and I would have fathers seriously reflect, and to be well assured, that the way to make their daughters to be long admired, beloved and respected by their husbands, is to make them skilful, able and active

in the most necessary concerns of a family. Eating and drinking come three times every day; the preparations for these, and all the ministry necessary to them, belong to the wife; and I hold it to be impossible, that at the end of two years, a really ignorant, sluttish wife, should possess any thing worthy of the name of love from her husband. This, therefore, is a matter of far greater moment to the father of a family than, whether the Parson of the parish or the Methodist Priest, be the most " Evangelical " of the two; for it is here a question of the daughter's happiness or misery for life. And I have no hesitation to say, that if I were a labouring man I should prefer teaching my daughters to bake, brew, milk, make butter and cheese, to teaching them to read the Bible till they had got every word of it by heart; and I should think, too, nay, I should know, that I was in the former case doing my duty towards God as well as towards my children.

89. When we see a family of dirty ragged little creatures, let us inquire into the cause, and ninety-nine times out of every hundred we shall find that the parents themselves have been brought up in the same way. But a consideration which ought of itself to be sufficient, is the contempt in which a husband will naturally hold a wife that is ignorant of the matters necessary to the conducting of a family. A woman who understands all the things above mentioned is really a skilful person; a person worthy of respect, and that will be treated with respect, too, by all the brutish employers or brutish husbands; and such, though sometimes, are not very frequently

found. Besides, if natural justice and our own interests had not the weight which they have, such valuable persons will be treated with respect. They know their own worth, and accordingly they are more careful of their character, more careful not to lessen by misconduct the value which they possess from their skill and ability.

90. Thus then the interests of the labourer, his health, the health of his family, the peace and happiness of his home, the prospects of his children through life, their skill, their ability, their habits of cleanliness, and even their moral deportment; all combine to press upon him the adoption and the constant practice of this branch of domestic economy. " Can she bake? " is the question that I always put. If she can, she is worth a pound or two a year more. Is that nothing? Is it nothing for a labouring man to make his four or five daughters worth eight or ten pounds a year more; and that too while he is by the same means providing the more plentifully for himself and the rest of his family? The reasons are the side of the thing that I contend for are endless; but if this one motive be not sufficient, I am sure all that I have said, and all that I could say, must be wholly unavailing.

91. Before, however, I dismiss this subject, let me say a word or two to those persons who do not come under the denomination of labourers. In London, or in any very large town where the space is so confined, and where the proper fuel is not handily to be come at and stored for use, to bake your own bread may be attended with too much

difficulty; but in all other situations there appears
to me to be hardly any excuse for not making bread
at home. If the family consist of twelve or fourteen
persons, the money actually saved in this way (even
at present prices) would be little short of from twenty
to thirty pounds a year. At the utmost here is only
the time of one woman occupied one day in the
week. Now mind, here are twenty-five pounds to be
employed in some way different from that of giving
it to the baker. If you add five of these pounds to a
woman's wages, is not that full as well employed as
giving it in wages to the baker's men? Is it not better
employed for you? and is it not better employed for
the community? It is very certain, that if the practice
were as prevalent as I could wish, there would be a
large deduction from the regular baking population;
but would there be any harm if less alum were imported into England, and if some of those youths
were left at the plough, who are now bound in
apprenticeships to learn the art and mystery of doing
that which every girl in the kingdom ought to be
taught to do by her mother? It ought to be a maxim
with every master and every mistress never to employ
another to do that which can be done as well by
their own servants. The more of their money that is
retained in the hands of their own people, the better
it is for them altogether. Besides, a man of right
mind must be pleased with the reflection, that there
is a great mass of skill and ability under his own
roof. He feels stronger and more independent on
this account, all pecuniary advantage out of the
question. It is impossible to conceive anything more

contemptible than a crowd of men and women living together in a house, and constantly looking out of it for people to bring them food and drink, and to fetch their garments to and fro. Such a crowd resemble a nest of unfledged birds absolutely dependent for their existence on the activity and success of the old ones.

92. Yet on men go, from year to year, in this state of wretched dependence, even when they have all the means of living within themselves, which is certainly the happiest state of life that any one can enjoy. It may be asked, Where is the mill to be found? where is the wheat to be got? The answer is, Where is there not a mill? where is there not a market? They are everywhere, and the difficulty is to discover what can be the particular attractions contained in that long and luminous manuscript, a baker's half-yearly bill.

93. With regard to the mill, in speaking of families of any considerable number of persons, the mill has, with me, been more than once a subject of observation in print. I for a good while experienced the great inconvenience and expense of sending my wheat and other grain to be ground at a mill. This expense, in case of a considerable family, living at only a mile from a mill, is something; but the inconveniency and uncertainty are great. In my " Year's Residence in America," from paragraphs 1031 and onwards, I give an account of a horse-mill which I had in my farmyard; and I showed, I think very clearly, that corn could be ground cheaper in this way than by wind or water, and that it would answer well to

grind for sale in this way as well as for home use. Since my return to England I have seen a mill, erected in consequence of what the owner had read in my book. This mill belongs to a small farmer, who, when he cannot work on his land with his horses, or in the season when he has little for them to do, grinds wheat, sells the flour, and he takes in grists to grind, as other millers do. This mill goes with three small horses; but what I would recommend to gentlemen with considerable families, or to farmers, is a mill such as I myself have at present.

94. With this mill, turned by a man and a stout boy, I can grind six bushels of wheat in a day and dress the flour. The grinding of six bushels of wheat at ninepence a bushel comes to four and sixpence, which pays the man and the boy, supposing them (which is not and seldom can be the case) to be hired for the express purpose out of the street. With the same mill you grind meat for your pigs; and of this you will get eight or ten bushels ground in a day. You have no trouble about sending to the mill; you are sure to have your own wheat; for strange as it may seem, I used sometimes to find that I sent white Essex wheat to the mill, and that it brought me flour from very coarse red wheat. There is no accounting for this, except by supposing that wind and water power has something in it to change the very nature of the grain; as when I came to grind by horses, such as the wheat went into the hopper, so the flour came out into the bin.

95. But mine now is only on the petty scale of providing for a dozen of persons and a small lot of

pigs. For a farm house, or a gentleman's house in the country, where there would be room to have a walk for a horse, you might take the labour from the men, clap any little horse, pony, or even ass, to the wheel; and he would grind you off eight or ten bushels of wheat in a day, and both he and you would have the thanks of your men into the bargain.

96. The cost of this mill is twenty pounds. The dresser is four more; the horse-path and wheel might possibly be four or five more; and I am very certain that to any farmer living at a mile from the mill (and that is less than the average distance perhaps), having twelve persons in family, having forty pigs to feed, and twenty hogs to fatten, the savings of such a mill would pay the whole expenses of it the very first year. Such a farmer cannot send less than fifty times a year to the mill. Think of that in the first place! The elements are not always propitious; sometimes the water fails and sometimes the wind. Many a farmer's wife has been tempted to vent her spleen on both. At best there must be horse and man, or boy, and perhaps cart, to go to the mill; and that too, observe, in all weathers, and in the harvest as well as at other times of the year. The case is one of imperious necessity; neither floods nor droughts, nor storms, nor calms, will allay the cravings of the kitchen, nor quiet the clamorous uproar of the stye. Go somebody must, to some place or other, and back they must come with flour and with meal. One summer many persons came down the country more than fifty miles to a mill that I knew in Pennsylvania; and I have known farmers

in England carry their grists more than fifteen miles to be ground. It is surprising, that, under these circumstances, hand-mills and horse-mills should not, long ago, have become of more general use; especially when one considers that the labour, in this case, would cost the farmer next to nothing. To grind would be the work of a wet day. There is no farmer who does not at least fifty days in every year exclaim, when he gets up in the morning, " What shall I set them at to-day? " If he had a mill, he would make them pull off their shoes, sweep all out clean, winnow up some corn, if he had it not already done, and grind and dress, and have everything in order. No scolding within doors about the grist, no squeaking in the stye, no boy sent off in the rain to the mill.

97. But there is one advantage which I have not yet mentioned, and which is the greatest of all; namely, that you would have the power of supplying your married labourers; your blacksmith's men sometimes; your wheelwright's men at other times; and, indeed, the greater part of the persons that you employed, with good flour, instead of their going to purchase their flour, after it had passed through the hands of a Corn Merchant, a Miller, a Flour Merchant, and a Huckster, every one of whom does and must have a profit out of the flour, arising from wheat grown upon, and sent away from, your very farm! I used to let all my people have flour at the same price that they would otherwise have been compelled to give for worse flour. Every farmer will understand me when I say, that he ought to pay for nothing in money, which he can pay for in anything

but money. His maxim is to keep the money that he takes as long as he can. Now here is a most effectual way of putting that maxim in practice to a very great extent. Farmers know well that it is the Saturday night which empties their pockets, and here is the means of cutting off a good half of the Saturday night. The men have better flour for the same money, and still the farmer keeps at home those profits which would go to the maintaining of the dealers in wheat and in flour.

98. The maker of my little mill is Mr. HILL, of Oxford-street. The expense is what I have stated it to be. I, with my small establishment, find the thing convenient and advantageous; what then must it be to a gentleman in the country who has room and horses, and a considerable family to provide for? The dresser is so contrived as to give you, at once, meal of four degrees of fineness; so that, for certain purposes, you may take the very finest; and, indeed, you may have your flour, and your bread of course, of what degree of fineness you please. But there is also a steel-mill, much less expensive, requiring less labour, and yet quite sufficient for a family. Mills of this sort, very good and at a reasonable price, are to be had of Mr. PARKES, in Fenchurch-street, London. These are very complete things of their kind. Mr. PARKES has, also, excellent malt-mills.

99. In concluding this part of my Treatise, I cannot help expressing my hope of being instrumental in inducing a part of the labourers, at any rate, to bake their own bread; and, above all things, to abandon the use of " Ireland's *lazy* root." Never-

theless, so extensive is the erroneous opinion relative
to this villainous root, that I really began to despair
of checking its cultivation and use, till I saw the
declaration which Mr. WAKEFIELD had the good
sense and the spirit to make before the " AGRICUL-
TURAL COMMITTEE." Be it observed, too, that Mr.
WAKEFIELD had himself made a survey of the state
of Ireland. What he saw there did not encourage
him, doubtless, to be an advocate for the growing of
this root of wretchedness. It is an undeniable fact,
that, in the proportion that this root is in use, as a
substitute for bread, the people are wretched; the
reasons for which I have explained and enforced a
hundred times over. Mr. WILLIAM HANNING told
the Committee, that the labourers in his part of
Somersetshire were " almost wholly supplied with
" potatoes, breakfast and dinner, brought them in the
" fields, and nothing but potatoes; and that they used
" in better times to get a certain portion of bacon
" and cheese, which, on account of their poverty,
" they do not eat now." It is impossible that men
can be contented in such a state of things: it is unjust
to desire them to be contented: it is a state of misery
and degradation to which no part of any community
can have any show of right to reduce another part:
men so degraded have no protection; and it is a dis-
grace to form part of a community to which they
belong. This degradation has been occasioned by a
silent change in the value of the money of the country. This has purloined the wages of the labourer;
it has reduced him by degrees to house with the
spider and the bat, and to feed with the pig. It has

changed the habits and in a great measure the character of the people. The sins of this system are enormous and undescribable; but, thank God! they seem to be approaching to their end! Money is resuming its value, labour is recovering its price; let us hope that the wretched potato is disappearing, and that we shall once more see the knife in the labourer's hand and the loaf upon his board.

[This was written in 1821. Now (1823) we have had the experience of 1822, when for the first time the world saw a considerable part of the people plunged into all the horrors of famine, at a moment when the government of that nation declared food to be abundant. Yes, the year 1822 saw Ireland in this state; saw the people of whole parishes receiving the extreme unction preparatory to yielding up their breath for want of food; and this while large exports of meat and flour were taking place in that country! But horrible as this was, disgraceful as it was to the name of Ireland, it was attended with this good effect: it brought out, from many Members of Parliament (in their places), and from the public in general, the acknowledgment, that the misery and degradation of the Irish were chiefly owing to the use of the potato as the almost sole food of the people.]

100. In my next Number I shall treat of the keeping of cows. I have said that I will teach the Cottager how to keep a cow all the year round upon the produce of a quarter of an acre, or in other words, forty rods, of land; and in my next I will make good my promise.

No. IV

MAKING BREAD

(*Continued*)

101. IN the last Number, at Paragraph 86, I observed that I hoped it was unnecessary for me to give any directions as to the mere act of making bread. But several correspondents inform me that without these directions a conviction of the utility of baking bread at home is of no use to them. Therefore I shall here give those directions, receiving my instructions here from one, who, I thank God, does know how to perform this act.

102. Suppose the quantity be a bushel of flour. Put this flour into a trough that people have for the purpose, or it may be in a clean smooth tub of any shape if not too deep, and if sufficiently large. Make a pretty deep hole in the middle of this heap of flour. Take (for a bushel) a pint of good fresh yeast, mix it and stir it well up in a pint of soft water milk-warm. Pour this into the hole in the heap of flour. Then take a spoon and work it round the outside of this body of moisture so as to bring into that body, by degrees, flour enough to make it form a thin batter, which you must stir about well for a minute or two. Then take a handful of flour and scatter it thinly over the head of this batter, so as to hide it. Then cover the whole over with a cloth to keep it warm; and this covering, as well as the situation of the trough, as to distance from the fire, must depend

on the nature of the place and state of the weather as to heat and cold. When you perceive that the batter has risen enough to make cracks in the flour that you covered it over with, you begin to form the whole mass into dough, thus: you begin round the hole containing the batter, working the flour into the batter, and pouring in, as it is wanted to make the flour mix with the batter, soft water milk-warm, or milk, as hereafter to be mentioned. Before you begin this you scatter the salt over the heap at the rate of half a pound to a bushel of flour. When you have got the whole sufficiently moist, you knead it well. This is a grand part of the business; for, unless the dough be well worked, there will be little round lumps of flour in the loaves; and besides, the original batter, which is to give fermentation to the whole, will not be duly mixed. The dough must, therefore, be well worked. The fists must go heartily into it. It must be rolled over, pressed out, folded up and pressed out again, until it be completely mixed and formed into a stiff and tough dough. This is labour, mind. I have never quite liked baker's bread since I saw a great heavy fellow in a bakehouse in France, kneading bread with his naked feet! His feet looked very white to be sure; whether they were of that colour before he got into the trough I could not tell. God forbid that I should suspect that this is ever done in England! It is labour, but what is exercise other than labour? Let a young woman bake a bushel once a week, and she will do very well without phials and gallipots.

103. Thus then the dough is made. And when

made it is to be formed into a lump in the middle of the trough, and with a little dry flour thinly scattered over it, covered over again to be kept warm and to ferment; and in this state, if all be done rightly, it will not have to remain more than about 15 or 20 minutes.

104. In the mean while the oven is to be heated, and this is much more than half the art of the operation. When an oven is properly heated, can be known only by actual observation. Women who understand the matter know when the heat is right the moment they put their faces within a yard of the oven-mouth; and once or twice observing is enough for any person of common capacity. But this much may be said in the way of rule, that the fuel (I am supposing a brick oven) should be dry (not rotten) wood, and not mere brushwood, but rather fagot-sticks. If larger wood, it ought to be split up into sticks not more than two, or two and a half inches through. Brushwood that is strong, not green and not too old, if it be hard in its nature and has some sticks in it, may do. The woody parts of furze, or ling, will heat an oven very well. But the thing is, to have a lively and yet somewhat strong fire, so that the oven may be heated in about 15 minutes, and retain its heat sufficiently long.

105. The oven should be hot by the time that the dough, as mentioned in Paragraph 103, has remained in the lump about 20 minutes. When both are ready, take out the fire, and wipe the oven out clean, and, at nearly about the same moment, take the dough out upon the lid of the baking trough, or some proper

place, cut it up into pieces, and make it up into loaves, kneading it again into the separate parcels; and as you go on, shaking a little flour over your board, to prevent the dough from adhering to it. The loaves should be put into the oven as quickly as possible after they are formed; when in, the oven-lid or door should be fastened up very closely; and, if all be properly managed, loaves of about the size of quartern loaves will be sufficiently baked in about two hours. But they usually take down the lid and look at the bread, in order to see how it is going on.

106. And what is there worthy of the name of plague or trouble in all this? Here is no dirt, no filth, no rubbish, no litter, no slop, And pray what can be pleasanter to behold? Talk indeed of your panto-mimes and gaudy shows, your processions and in-stallations and coronations! Give me for a beautiful sight, a neat and smart woman, heating her oven and setting her bread! And if the bustle does make the sign of labour glisten on her brow, where is the man that would not kiss that off, rather than lick the plaster from the cheek of a duchess?

107. And what is the result? Why, good, whole-some food, sufficient for a considerable family for a week, prepared in three or four hours. To get this quantity of food, fit to be eaten, in the shape of pota-toes, how many fires! what a washing, what a boiling, what a peeling, what a slopping, and what a messing! The cottage everlastingly in a litter; the woman's hands everlastingly wet and dirty; the children grimed up to the eyes with dust fixed on by potato-starch; and ragged as colts, the poor mother's time

all being devoted to the everlasting boiling of the pot! Can any man, who knows anything of the labourer's life, deny this? And will, then, anybody, except the old shuffle-breeches band of the Quarterly Review, who have all their lives been moving from garret to garret, who have seldom seen the sun, and never the dew except in print; will anybody, except these men, say, that the people ought to be taught to use potatoes as a substitute for bread?

BREWING BEER

108. THIS matter has been fully treated of in the two last Numbers. But several correspondents wishing to fall upon some means of rendering the practice beneficial to those who are unable to purchase brewing utensils, have recommended the lending of them, or letting out, round a neighbourhood. Another correspondent has, therefore, pointed out to me an Act of Parliament which touches upon this subject; and, indeed, what of Excise Laws and Custom Laws and Combination Laws and Libel Laws, a human being in this country scarcely knows what he dares do or what he dares say. What father, for instance, would have imagined, that, having brewing utensils, which two men carry from house to house as easily as they can a basket, he dared not lend them to his son living in the next street, or at the street door? Yet such really is the law; for according to the Act 5th of the 22 and 23 of that honest and sincere gentle-

man Charles II, there is a penalty of £50 for lending or letting brewing utensils. However, it has a limit, that the penalty is confined to Cities, Corporate Towns, and Market Towns, WHERE THERE IS A PUBLIC BREWHOUSE. So that, in the first place, you may let, or lend, in any place where there is no public brewhouse; and in all towns not corporate or market, and in all villages, hamlets, and scattered places.

109. Another thing is, can a man who has brewed beer at his own house in the country bring that beer into town to his own house and for the use of his family there? This has been asked of me. I cannot give a positive answer without reading about seven large volumes in quarto of taxing laws. The best way would be to try it; and if any penalty, pay it by subscription, if that would not come under the law of conspiracy! However, I think there can be no danger here. So monstrous a thing as this can surely not exist. If there be such a law, it is daily violated; for nothing is more common than for country gentlemen, who have a dislike to die by poison, bringing their home-brewed beer to London.

110. Another correspondent recommends Parishes to make their own malt. But surely the landlords mean to get rid of the malt and salt tax! Many dairies, I dare say, pay £50 a year each in salt tax. How, then, are they to contend against Irish butter and Dutch butter and cheese? And as to the malt tax, it is a dreadful drain from the land. I have heard of labourers, living in "unkent places," making their own malt,

even now! Nothing is so easy as to make your own malt, if you were permitted. You soak the barley about three days (according to the state of the weather), and then you put it upon stones or bricks, and keep it turned, till the root shoots out; and then, to know when to stop, and to put it to dry, take up a corn (which you will find nearly transparent) and look through the skin of it. You will see the spear, that is to say, the shoot that would come out of the ground, pushing on towards the point of the barley-corn. It starts from the bottom, where the root comes out; and it goes on toward the other end; and would, if kept moist, come out at that other end when the root was about an inch long. So that, when you have got the root to start, by soaking and turning in heap, the spear is on its way. If you look in through the skin you will see it; and now observe, when the point of the spear has got along as far as the middle of the barley corn, you should take your barley and dry it. How easy would every family and especially every farmer, do this, if it were not for the punishment attached to it. The persons in the " unkent places " before mentioned dry the malt in their oven! But let us hope that the labourer will soon be able to get malt without exposing himself to punishment as a violator of the law.

KEEPING COWS

111. As to the use of milk and of that which proceeds from milk, in a family, very little need be said. At a certain age bread and milk are all that a child

wants. At a later age they furnish one meal a day for children. Milk is, at all seasons, good to drink. In the making of puddings, and in the making of bread too, how useful is it! Let any one who has eaten none but baker's bread for a good while taste bread home-baked, mixed with milk instead of with water, and he will find what the difference is. There is this only to be observed, that in hot weather bread mixed with milk will not keep so long as that mixed with water. It will of course turn sour sooner.

112. Whether the milk of a cow be to be consumed by a cottage family in the shape of milk, or whether it be to be made to yield butter, skim-milk, and butter-milk, must depend on circumstances. A woman that has no child, or only one, would, perhaps, find it best to make some butter at any rate. Besides, skim-milk and bread (the milk being boiled) is quite strong food enough for any children's breakfast, even when they begin to go to work; a fact which I state upon the most ample and satisfactory experience, very seldom having ever had any other sort of breakfast myself till I was more than ten years old, and I was in the fields at work full four years before that. I will here mention that it gave me singular pleasure to see a boy, just turned of six, helping his father to reap, in Sussex, this last summer. He did little, to be sure, but it was something. His father set him into the ridge at a great distance before him, and when he came up to the place he found a sheaf cut; and those who know what it is to reap, know how pleasant it is to find now and then a sheaf cut ready to their hand. It was no small thing to see

a boy fit to be trusted with so dangerous a thing as
a reap-hook in his hands, at an age when " young
masters " have nursery-maids to cut their victuals
for them, and to see that they do not fall out of the
window, tumble down stairs, or run under carriage-
wheels or horses' bellies. Was not this father dis-
charging his duty by this boy much better than he
would have been by sending him to a place called a
school? The boy is in a school here, and an excellent
school too: the school of useful labour. I must hear
a great deal more than I ever have heard, to convince
me that teaching children to read tends so much to
their happiness, their independence of spirit, their
manliness of character, as teaching them to reap.
The creature that is in want must be a slave; and to
be habituated to labour cheerfully is the only means
of preventing nineteen-twentieths of mankind from
being in want. I have digressed here; but observations of this sort can, in my opinion, never be too
often repeated; especially at a time when all sorts
of mad projects are on foot, for what is falsely called
educating the people, and when some would do this
by a tax that it would compel the single man to give
part of his earnings to teach the married man's
children to read and write.

113. Before I quit the use to which milk may be
put, let me mention, that, as mere drink, it is, unless
perhaps in case of heavy labour, better, in my opinion
than any beer, however good. I have drinked little
else for the last five years, at any time of the day.
Skim-milk I mean. If you have not milk enough to
wet up your bread with (for a bushel of flour requires

about 16 to 18 pints), you make up the quantity
with water of course; or, which is a very good way,
with water that has been put, boiling hot, upon
bran, and then drained off. This takes the goodness
out of the bran to be sure; but rather good bread is
a thing of so much importance that it always ought
to be the very first object in domestic economy.

114. The cases vary so much, that it is impossible
to lay down rules for the application of the produce
of a cow, which rules shall fit all cases. I content
myself, therefore, with what has already been said
on this subject; and shall only make an observation
on the act of milking, before I come to the chief
matter; namely, the getting of the food for the cow.
A cow should be milked clean. Not a drop, if it can
be avoided, should be left in the udder. It has been
proved that the half pint that comes out last has
twelve times, I think it is, as much butter in it, as
the half pint that comes out first. I tried the milk of
ten Alderney cows, and, as nearly as I, without being
very nice about the matter, could ascertain, I found
the difference to be about what I have stated. The
udder would seem to be a sort of milk-pan in which
the cream is uppermost, and of course comes out
last, seeing that the outlet is at the bottom. But
besides this, if you do not milk clean, the cow will
give less and less milk, and will become dry much
sooner than she ought. The cause of this I do not
know, but experience has long established the fact.

115. In providing food for a cow, we must look,
first, at the sort of cow; seeing that a cow of one sort
will certainly require more than twice as much food

as a cow of another sort. For a cottage, a cow of the smallest sort common in England is, on every account, the best; and such a cow will not require above 70 or 80 pounds of good moist food in the twenty-four hours.

116. Now, how to raise this food on 40 rods of ground is what we want to know. It frequently happens that a labourer has more than 40 rods of ground. It more frequently happens, that he has some common, some lane, some little outlet or other, for a part of the year, at least. In such case he may make a different disposition of his ground, or may do with less than the 40 rods. I am here, for simplicity's sake, to suppose, that he have 40 rods of clear, unshaded land, besides what his house and sheds stand upon; and that he have nothing further in the way of means to keep his cow.

117. I suppose the 40 rods to be clean and unshaded: for I am to suppose that when a man thinks of 5 quarts of milk a day, on the average, all the year round, he will not suffer his ground to be encumbered by apple-trees that give only the means of treating his children to fits of the belly-ache, or with currant and gooseberry bushes, which, though their fruit do very well to amuse, really give nothing worthy of the name of food, except to the black-birds and thrushes. The ground is to be clear of trees; and in the spring we will suppose it to be clean. Then dig it up deeply, or, which is better, trench it, keeping, however, the top spit of the soil at the top. Lay it in ridges, in April or May, about two feet apart, and made high and sharp. When the weeds appear about

three inches high, turn the ridges into the furrows (never moving the ground but in dry weather), and bury all the weeds. Do this as often as the weeds get three inches high, and by the fall you will have really clean ground and not poor ground.

118. There is the ground then ready. About the 26th of August, but not earlier, prepare a rod of your ground, and put some manure in it (for some you must have), and sow one half of it with Early York Cabbage Seed, and the other half with Sugar-loaf Cabbage Seed, both of the true sort, in little drills, at 8 inches apart, and the seeds thin in the drill. If the plants come up at two inches apart (and they should be thinned if thicker), you will have a plenty. As soon as fairly out of the ground, hoe the ground nicely, and pretty deeply, and again in a few days. When the plants have six leaves, which will be very soon, dig up, make fine, and manure another rod or two, and prick out the plants, 4,000 of each, in rows at eight inches apart and three inches in the row. Hoe the ground between them often, and they will grow fast and be straight and strong. I suppose that these beds for plants take 4 rods of your ground. Early in November, or, as the weather may serve, a little earlier or later, lay some manure (of which I shall say more hereafter) between the ridges, in the other 36 rods, and turn the ridges over on this manure, and then transplant your plants on the ridges, at 15 inches apart. Here they will stand the winter; and you must see that the slugs do not eat them. If your plants fail, you have plenty in the bed where you prick them out; for your 36 rods

will not require more than 4,000 plants. If the
winter be very hard, and bad for plants, you cannot
cover 36 rods, but you may the bed where the rest
of your plants are. A little litter, or straw, or dead
grass, or fern, laid along between the rows and the
plants, not to cover the leaves, will preserve them
completely.   When people complain of all their
plants being " cut off," they have, in fact, nothing
to complain of but their own extreme carelessness.
If I had a gardener who complained of all his plants
being cut off, I should cut him off pretty quickly.
If those in the 36 rods fail, or fail in part, fill up their
places, later in the winter, by plants from the bed.

119. If you find the ground dry at top during the
winter, hoe it, and particularly near the plants, and
root out all slugs and insects. And when March
comes, and the ground is dry, hoe deep and well,
and earth the plants up close to the lower leaves.
As soon as the plants begin to grow, dig the ground
with a spade clean and well, and let the spade go as
near to the plants as you can without actually dis-
placing the plants. Give them another digging in a
month; and, if weeds come in the meanwhile, hoe,
and let not one live a week. " Oh! what a deal of
work! " Well! but it is for yourself, and, besides, it
is not all to be done in a day; and we shall by-and-
by see what it is altogether.

120. By the first of June; I speak of the South of
England, and there is also some difference in seasons
and soils; but, generally speaking, by the first of
June, you will have turned-in-cabbages, and soon
you will have the Early Yorks solid. And by the first

of June you may get your cow, one that is about to calve, or that has just calved, and at this time such a cow as you will want will not, thank God, cost above five pounds.

121. I shall speak of the place to keep her in, and of the manure and litter by-and-by. At present I confine myself to her mere food. The 36 rods, if the cabbages all stood till they got solid, would give her food for 200 days, at 80 pounds weight per day, which is more than she would eat. But you must use some, at first, that are not solid; and then, some of them will split before you can use them. But you will have pigs to help off with them, and to gnaw the heads of the stumps. Some of the Sugar-loaves may have been planted out in the spring; and thus these 36 rods will get you along to some time in September.

122. Now mind, in March, and again in April, sow more Early Yorks, and get them to be fine stout plants, as you did those in the fall. Dig up the ground and manure it, and, as fast as you cut cabbages, plant cabbages; and in the same manner and with the same cultivation as before. Your last planting will be about the middle of August, with stout plants, and these will serve you into the month of November.

123. Now we have to provide from December to May inclusive; and that, too, out of this same piece of ground. In November there must be, arrived at perfection, 3,000 turnip plants. These, without the greens, must weigh, on an average, 5 pounds, and this, at 80 pounds a day, will keep the cow 187 days; and there are but 182 days in these six months. The

greens will have helped out the latest cabbages to carry you through November, and perhaps into December. But for these six months, you must depend on nothing but the Swedish turnips.

124. And now how are these to be had upon the same ground that bears the cabbages? That we are now going to see. When you plant out your cabbages at the outset, put first a row of Early Yorks, then a row of Sugar-loaves, and so on throughout the piece. Of course, as you are to use the Early Yorks first, you will cut every other row; and the Early Yorks that you are to plant in summer will go into the intervals. By-and-by the Sugar-loaves are cut away, and in their place will come Swedish turnips, you digging and manuring the ground as in the case of the cabbages; and at last you will find about 16 rods where you will have found it too late, and unnecessary besides, to plant any second crop of cabbages. Here the Swedish turnips will stand in rows at two feet apart (and always a foot apart in the row), and thus you will have three thousand turnips; and if these do not weigh five pounds each on an average, the fault must be in the seed or in the management.

125. The Swedish turnips are raised in this manner. You will bear in mind the four rods of ground in which you have sowed and pricked out your cabbage plants. The plants that will be left there will, in April, serve you for greens, if you ever eat any, though bread and bacon are very good without greens, and rather better than with. At any rate, the pig, which has strong powers of digestion, will consume this herbage. In a part of these four

rods you will, in March and April, as before directed, have sown and raised your Early Yorks for the summer planting. Now, in the last week of May, prepare a quarter of a rod of this ground, and sow it, precisely as directed for the Cabbage-seed, with Swedish turnip seed, and sow a quarter of a rod every three days, till you have sowed two rods. If the fly appear, cover the rows over in the day-time with cabbage leaves, and take the leaves off at night; hoe well between the plants; and when they are safe from the fly, thin them to four inches apart in the row. The two rods will give you nearly five thousand plants, which is two thousand more than you will want. From this bed you draw your plants to transplant in the ground where the cabbages have stood, as before directed. You should transplant none much before the middle of July, and not much later than the middle of August. In the two rods, whence you take your turnip plants, you may leave plants to come to perfection, at two-feet distances each way; and this will give you, over and above, 840 pounds weight of turnips. For the other two rods will be ground enough for you to sow your cabbage plants in at the end of August, as directed for last year.

126. I should now proceed to speak of the manner of harvesting, preserving, and using the crops; of the manner of feeding the cow; of the shed for her; of the managing of the manure, and several other less important things; but these, for want of room here, must be reserved for the beginning of my next number. After, therefore, observing that the Turnip

plants must be transplanted in the same way that Cabbage plants are, and that both ought to be transplanted in dry weather and in ground just fresh digged, I shall close this number with the notice of two points, which I am most anxious to impress upon the mind of every reader.

127. The first is, whether these crops give an ill taste to milk and butter. It is very certain, that the taste and smell of certain sorts of cattle-food will do this; for, in some parts of America, where the wild garlick, of which the cows are very fond, and which, like other bulbous-rooted plants, spring before the grass, not only the milk and butter have a strong taste of garlick, but even the veal, when the calves suck milk from such sources. None can be more common expressions, than, in Philadelphia market, are those of Garlicky Butter and Garlicky Veal. I have distinctly tasted the Whiskey in milk of cows fed on distillers' wash. It is also certain, that, if the cow eat putrid leaves of cabbages and turnips, the butter will be offensive. And the white turnip, which is at best but a poor thing and often half putrid, makes miserable butter. The large cattle cabbage, which, when loaved hard, has a strong and even an offensive smell, will give a bad taste and smell to butter, whether there be putrid leaves or not. If you boil one of these rank cabbages, the water is extremely offensive to the smell. But I state upon positive and recent experience, that Early York and Sugar-loaf Cabbages will yield as sweet milk and butter as any food that can be given to a cow. During this last summer, I have, with the exception about

to be noticed, kept, from the 1st of May to the 22nd of October, five cows upon the grass of two acres and a quarter of ground, the grass being generally cut up for them and given to them in the stall. I had in the spring 5,000 cabbage plants, intended for my pigs, eleven in number. But the pigs could not eat half their allowance, though they were not very small when they began upon it. We were compelled to resort to the aid of the cows; and, in order to see the effect on the milk and butter, we did not mix the food; but gave the cows two distinct spells at the cabbages, each spell about ten days in duration. The cabbages were cut off the stump with little or no care about dead leaves. And sweeter, finer butter, butter of a finer colour, than these cabbages made, never was made in this world. I never had better from cows feeding in the sweetest pasture. Now, as to Swedish turnips, they do give a little taste, especially if boiling of the milk pans be neglected, and if the greatest care be not taken about all the dairy tackle. Yet we have, for months together, had the butter so fine from Swedish turnips, that nobody could well distinguish it from grass-butter. But to secure this, there must be no sluttishness. Churn, pans, pail, shelves, wall, floor, and all about the dairy, must be clean; and, above all things, the pans must be boiled. However, after all, it is not here a case of delicacy of smell so refined as to faint at anything that meets it except the stink of perfumes. If the butter do taste a little of the Swedish turnip, it will do very well where there is plenty of that sweet sauce which early rising and bodily labour are ever sure to bring.

128. The other point (about which I am still more anxious) is the seed; for if the seed be not sound, and especially if it be not true to its kind, all your labour is in vain. It is best, if you can do it, to get your seed from some friend, or some one that you know and can trust. If you save seed, observe all the precautions mentioned in my book on Gardening. This very year I have some Swedish turnips, so called, about 7,000 in number, and should, if my seed had been true, have had about twenty tons weight, instead of which I have about three! Indeed, they are not Swedish turnips, but a sort of mixture between that plant and rape. I am sure the seedsman did not wilfully deceive me. He was deceived himself. The truth is, that seedsmen are compelled to buy their seeds of this plant. Farmers save it; and they but too often pay very little attention to the manner of doing it. The best way is to get a dozen of fine turnip plants, perfect in all respects, and plant them in a situation where the smell of the blossoms of nothing of the cabbage or rape or turnip or even charlock kind, can reach them. The seed will keep perfectly good for four years.

No. V

KEEPING COWS

(*Continued*)

129. I HAVE now, in the conclusion of this article, to speak of the manner of harvesting and preserving the Swedes; of the place to keep the cow in; of the

manure for the land: and of the quantity of labour that the cultivation of the land and the harvesting of the crop will require.

130. Harvesting and preserving the Swedes. When they are ready to take up, the tops must be cut off, if not cut off before, and also the roots; but neither tops nor roots should be cut off very close. You will have room for ten bushels of the bulbs in the house, or shed. Put the rest into ten-bushel heaps. Make the heap upon the ground in a round form, and let it rise up to a point. Lay over it a little litter, straw, or dead grass, about three inches thick, and then earth upon that about six inches thick. Then cut a thin round green turf, about eighteen inches over, and put it upon the crown of the heap to prevent the earth from being washed off. Thus these heaps will remain till wanted for use. When given to the cow, it will be best to wash the Swedes and cut each into two or three pieces with a spade or some other tool. You can take in ten bushels at a time. If you find them sprouting in the spring, open the remaining heaps, and expose them to the sun and wind; and cover them again slightly with straw or litter of some sort.*

131. As to the place to keep the cow in, much will depend upon situation and circumstances. I am always supposing that the cottage is a real cottage, and not a house in a town or village street; though,

---

* Be sure, now, before you go any further, to go to the end of the book, and there read about MANGEL WURZEL. Be sure to do this. And there read also about COBBETT'S CORN. Be sure to do this before you go any further.

wherever there is the quarter of an acre of ground, the cow may be kept. Let me, however, suppose that which will generally happen; namely, that the cottage stands by the side of a road, or lane, and amongst fields and woods, if not on the side of a common. To pretend to tell a country labourer how to build a shed for a cow, how to stick it up against the end of the house, or to make it an independent erection; or to dwell on the materials, where poles, rods, wattles, rushes, furze, heath, and cooper-chips, are all to be gotten by him for nothing or next to nothing, would be useless: because a man who, thus situated, can be at any loss for a shed for his cow, is not only unfit to keep a cow, but unfit to keep a cat. The warmer the shed is, the better it is. The floor should slope, but not too much. There are stones, of some sort or other, everywhere, and about six wheel-barrows full will pave the shed, a thing to be by no means neglected. A broad trough, or box, fixed up at the head of the cow, is the thing to give her food in; and she should be fed three times a day, at least; always at day-light and at sun-set. It is not absolutely necessary that a cow ever quit her shed, except just at calving time, or when taken to the bull. In the former case the time is, nine times out of ten, known to within forty-eight hours. Any enclosed field or place will do for her during a day or two; and for such purpose, if there be not room at home, no man will refuse place for her in a fallow field. It will, however, be good, where there is no common to turn her out upon, to have her led by a string, two or three times a week, which may be done by a child only five years old, to graze, or pick, along the sides of roads and lanes. Where there is a common, she will, of course, be turned out in the day-time, except in very wet or severe weather; and in a case like this, a smaller quantity of ground will suffice for the keeping of her. According to the present practice, a miserable " tallet " of bad hay, is, in such cases, the winter provision for the cow. It can scarcely be called food; and the consequence is, the cow is both dry and lousy nearly half the year; instead of being dry only about fifteen days before calving, and being sleek and lusty at the end of the winter, to which a warm lodging greatly contributes. For, observe, if you keep a cow, any time between September and June, out in a field, or yard, to endure the chances of the weather, she will not, though she have food precisely the same in quantity and quality, yield above two-thirds as much as if she were lodged in house: and in wet weather she will not yield half so much. It is not so much the cold as the wet that is injurious to all out stock in England.

132. The Manure. At the beginning this must be provided by collections made on the road; by the results of the residence in a cottage. Let any man clean out every place about his dwelling; rake and scrape and sweep all into a heap; and he will find that he has a great deal. Earth of almost any sort that has long lain on the surface, and has been trodden on, is a species of manure. Every act that tends to neatness round a dwelling, tends to the creating of a mass of manure. And I have very seldom seen a cottage, with a plat of ground of a quarter of an acre

belonging to it, round about which I could not have
collected a very large heap of manure. Every thing
of animal or vegetable substance that comes into a
house, must go out of it again, in one shape or another.
The very emptying of vessels of various kinds, on a
heap of common earth, makes it a heap of the best
manure. Thus goes on the work of reproduction;
and thus is verified the words of the Scripture, " Flesh
is grass," and there is " Nothing new under the sun."
Thus far as to the outset. When you have got the cow,
there is no more care about manure; for, and especially
if you have a pig also, you must have enough annually
for an acre of ground. And let it be observed, that,
after a time, it will be unnecessary, and would be
injurious, to manure for every crop; for that would
produce more stalk and green than substantial part;
as it is well known, that wheat plants, standing in
ground too full of manure, will yield very thick and
long straws, but grains of little or no substance.
You ought to depend more on the spade and the
hoe than on the dung-heap. Nevertheless, the greatest
care should be taken to preserve the manure; because you will want straw, unless you be by the side
of a common which gives you rushes, grassy furze,
or fern; and to get straw you must give a part of
your dung from the cow-stall and pig-stye. The
best way to preserve manure, is to have a pit of
sufficient dimensions close behind the cow-shed
and pig-stye, for the run from these to go into, and
from which all runs of rain water should be kept.
Into this pit would go the emptying of the shed and
of the stye, and the produce of all sweepings and

cleanings round the house; and thus a large mass of manure would soon grow together: much too large a quantity for a quarter of an acre of ground. One good load of wheat or rye straw is all that you would want for the winter, and half of one for the summer; and you would have more than enough dung for the exchange against this straw.

133. Now, as to the quantity of labour that the cultivation of the land will demand in a year. We will suppose the whole to have five complete diggings, and say nothing about the little matters of sowing and planting, and hoeing and harvesting, all which are a mere trifle. We are supposing the owner to be an able labouring man; and such a man will dig 12 rods of ground in a day. Here are 200 rods to be digged, and here are little less than 17 days of work at 12 hours in the day; or 200 hours' work, to be done in the course of the long days of spring, and summer, while it is light long before six in the morning, and long after six at night. What is it then? Is it not better than time spent in the ale-house, or in creeping about after a miserable hare! Frequently, and most frequently, there will be a boy, if not two, big enough to help. And (I only give this as a hint) I saw, on the 7th of November last (1822), *a very pretty woman*, in the village of Hannington, in Wiltshire, digging a piece of ground and planting it with Early Cabbages, which she did as handily and as neatly as any gardener that ever I saw. The ground was wet, and therefore, to avoid treading the digged ground in that state, she had her line extended, and put in the rows as she advanced in

her digging, standing in the trench while she performed the act of planting, which she did with great nimbleness and precision. Nothing could be more skilfully or beautifully done. Her clothes were neat, clean, and tight about her. She had turned her handkerchief down from her neck, which, with the glow that the work had brought into her cheeks, formed an object which I do not say would have made me actually stop my chaise had it not been for the occupation in which she was engaged; but, all taken together, the temptation was too strong to be resisted. But there is the Sunday; and I know of no law, human or divine, that forbids a labouring man to dig or plant his garden on Sunday, if the good of his family demand it; and if he cannot, without injury to that family, find other time to do it in. Shepherds, carters, pigfeeders, drovers, coachmen, cooks, footmen, printers, and numerous others, work on the Sundays. Theirs are deemed by the law works of necessity. Harvesting and hay making are allowed to be carried on on the Sunday, in certain cases; when they are always carried on by provident farmers. And I should be glad to know the case which is more a case of necessity than that now under our view. In fact, the labouring people do work on the Sunday morning in particular, all over the country, at something or other, or they are engaged in pursuits a good deal less religious than that of digging and planting. So that, as to the 200 hours, they are easily found, without the loss of any of the time required for constant daily labour.

134. And what a produce is that of a cow! I

suppose only an average of 5 quarts of milk a day.
If made into butter, it will be equal every week to
two days of the man's wages, besides the value of
the skim milk; and this can hardly be of less value
than another day's wages. What a thing, then, is this
cow, if she earn half as much as the man! I am greatly
under-rating her produce: but I wish to put all the
advantages at the lowest. To be sure, there is work
for the wife, or daughter, to milk and make butter.
But the former is done at the two ends of the day,
and the latter only about once in the week. And,
whatever these may subtract from the labours of the
field, which all country women ought to be engaged
in whenever they conveniently can; whatever the
cares created by the cow may subtract from these,
is amply compensated for by the education that
these cares will give to the children. They will all
learn to milk,* and the girls to make butter. And,
which is a thing of the very first importance, they
will all learn, from their infancy, to set a just value
upon dumb animals, and will grow up in the habit
of treating them with gentleness and feeding them
with care. To those who have not been brought up

---

* To me the following has happened within the last year.
A young man, in the country, had agreed to be my servant;
but it was found that he could not milk; and the bargain was
set aside. About a month afterwards a young man, who said
he was a farmer's son, and who came from Herefordshire,
offered himself to me at Kensington. "Can you milk?" He
could not; but would learn! Ay, but in the learning he might
dry up my cows! What a shame to the parents of these young
men! Both of them were in want of employment. The latter
had come more than a hundred miles in search of work; and
here he was left to hunger still, and to be exposed to all sorts
of ills, because he could not milk.

in the midst of rural affairs, it is hardly possible to give an adequate idea of the importance of this part of education. I should be very loath to intrust the care of my horses, cattle, sheep, or pigs, to any one whose father never had cow or pig of his own. It is a general complaint, that servants, and especially farm-servants, are not so good as they used to be. How should they? They were formerly the sons and daughters of small farmers; they are now the progeny of miserable propertyless labourers. They have never seen an animal in which they had any interest. They are careless by habit. This monstrous evil has arisen from causes which I have a thousand times described; and which causes must now be speedily removed, or, they will produce a dissolution of society, and give us a beginning afresh.

135. The circumstances vary so much, that it is impossible to lay down precise rules suited to all cases. The cottage may be on the side of a forest or common; it may be on the side of a lane or of a great road, distant from town or village; it may be on the skirts of one of these latter; and then, again, the family may be few or great in number, the children small or big, according to all which circumstances, the extent and application of the cow-food, and also the application of the produce, will naturally be regulated. Under some circumstances, half the above crop may be enough; especially where good commons are at hand. Sometimes it may be the best way to sell the calf as soon as calved; at others, to fat it; and, at others, if you cannot sell it, which sometimes happens, to knock it on the head

as soon as calved; for, where there is a family of small children, the price of a calf of two months old cannot be equal to the half of the value of the two months' milk. It is pure weakness to call it " a pity." It is a much greater pity to see hungry children crying for the milk that a calf is sucking to no useful purpose; and as to the cow and the calf, the one must lose her young, and the other its life, after all; and the respite only makes an addition to the sufferings of both.

136. As to the pretended unwholesomeness of milk, in certain cases; as to its not being adapted to some constitutions, I do not believe one word of the matter. When we talk of the fruits, indeed, which were formerly the chief food of a great part of mankind, we should recollect that those fruits grew in countries that had a sun to ripen the fruits, and to put nutritious matter into them. But as to milk, England yields to no country upon the face of the earth. Neat cattle will touch nothing that is not wholesome in its nature; nothing that is not wholly innoxious. Out of a pail that has ever had grease in it, they will not drink a drop, though they be raging with thirst. Their very breath is fragrance. And how, then, is it possible that unwholesomeness should distil from the udder of a cow? The milk varies, indeed, in its quality and taste, according to the variations in the nature of the food; but no food will a cow touch that is in any way hostile to health. Feed young puppies upon milk from the cow, and they will never die with that ravaging disease called " the distemper." In short, to suppose that milk contains

any thing essentially unwholesome is monstrous.
When, indeed, the appetite becomes vitiated; when
the organs have been long accustomed to food of a
more stimulating nature; when it has been resolved
to eat ragouts at dinner, and drink wine, and to
swallow " a devil," and a glass of strong grog at
night; then milk for breakfast may be " heavy " and
disgusting, and the feeder may stand in need of tea
or laudanum, which differ only as to degrees of
strength. But, and I speak from the most ample
experience, milk is not " heavy," and much less is
it unwholesome, when he who uses it rises early,
never swallows strong drink, and never stuffs himself with flesh of any kind. Many and many a day
I scarcely taste of meat, and then chiefly at breakfast, and that, too, at an early hour. Milk is the
natural food of young people; if it be too rich, skim
it again and again till it be not too rich. This is an
evil easily cured. If you have now to begin with a
family of children, they may not like it at first. But
persevere; and the parent who does not do this,
having the means in his hands, shamefully neglects
his duty. A son who prefers a " devil " and a glass
of grog to a hunch of bread and a bowl of cold milk,
I regard as a pest, and for this pest the father has
to thank himself.

137. Before I dismiss this article, let me offer an
observation or two to those persons who live in the
vicinity of towns, or in towns, and who, though they
have large gardens, have " no land to keep a cow,"
a circumstance which they " exceedingly regret."
I have, I dare say, witnessed this case at least a thou-

sand times. Now, how much garden ground does it
require to supply even a large family with garden
vegetables? The market gardeners round the metrop-
olis of this wen-headed country; round the Wen of
all wens; round this prodigious and monstrous
collection of human beings; these market gardeners
have about three hundred thousand families to
supply with vegetables, and these they supply well
too, and with summer fruits into the bargain. Now,
if it demanded ten rods to a family, the whole would
demand, all but a fraction, nineteen thousand acres
of garden ground. We have only to cast our eyes
over what there is, to know that there is not a fourth
of that quantity. A square mile contains, leaving
out parts of a hundred, 700 acres of land; and 19,000
acres occupy more than twenty-two square miles.
Are there twenty-two square miles covered with
the Wen's market gardens? The very question is
absurd. The whole of the market gardens from
Brompton to Hammersmith, extending to Battersea
Rise on the one side, and to the Bayswater Road on
the other side, and leaving out roads, lanes, nurseries,
pastures, corn-fields, and pleasure-grounds, do not,
in my opinion, cover one square mile. To the north
and south of the Wen there is very little in the way
of market garden; and, if on both sides of the Thames,
to the eastward of the Wen, there be three square
miles actually covered with market gardens, that is
the full extent. How, then, could the Wen be supplied, if it required ten rods to each family? To be
sure, potatoes, carrots and turnips, and especially
the first of these, are brought, for the use of the

Wen, from a great distance, in many cases. But so they are for the use of the persons I am speaking of; for a gentleman thinks no more of raising a large quantity of these things in his garden, than he thinks of raising wheat there. How is it, then, that it requires half an acre, or eighty rods, in a private garden, to supply a family, while these market gardeners supply all these families (and so amply too) from ten, or more likely, five, rods of ground to a family? I have shown, in the last Number, that nearly fifteen tons of vegetables can be raised in a year on forty rods of ground; that is to say, ten loads for a wagon and four good horses. And is not a fourth, or even an eighth, part of this weight, sufficient to go down the throats of a family in a year? Nay, allow that only a ton goes to a family in a year, it is more than six pounds weight a day; and what sort of a family must that be that really swallows six pounds weight a day? and this a market gardener will raise for them upon less than three rods of ground; for he will raise, in the course of the year, even more than fifteen tons upon forty rods of ground. What is it, then, that they do with the eighty rods of ground in a private garden? Why, in the first place, they have one crop where they ought to have three. Then they do not half till the ground. Then they grow things that are not wanted. Plant cabbages and other things, let them stand till they be good for nothing, and then wheel them to the rubbish heap. Raise as many radishes, lettuces, and as much endive, and as many kidney-beans, as would serve for ten families; and finally throw nine-tenths

of them away. I once saw not less than three rods of
ground, in a garden of this sort, with lettuces all
bearing seed. Seed enough for half a county. They
cut a cabbage here, and a cabbage there, and so let
the whole of the piece of ground remain undug, till
the last cabbage be cut. But after all, the produce,
even in this way, is so great that it never could be
gotten rid of, if the main part were not thrown away.
The rubbish heap always receives four-fifths even
of the eatable part of the produce.

138. It is not thus that the market gardeners proceed. Their rubbish heap consists of little besides
mere cabbage-stumps. No sooner is one crop on
the ground than they settle in their minds what is
to follow it. They clear as they go in taking off a
crop, and, as they clear they dig and plant. The
ground is never without seed in it, or plants on it.
And thus, in the course of the year, they raise a prodigious bulk of vegetables from eighty rods of ground.
Such vigilance and industry are not to be expected
in a servant; for it is foolish to expect that a man
will exert himself for another as much as he will for
himself. But if I was situated as one of the persons
is that I have spoken of in paragraph 137; that is to
say, if I had a garden of eighty rods, or even of sixty
rods of ground, I would, out of that garden, draw a
sufficiency of vegetables for my family, and would
make it yield enough for a cow besides. I should go
a short way to work with my gardener. I should put
" Cottage Economy " into his hands, and tell him,
that if he could furnish me with vegetables, and
my cow with food, he was my man; and that if he

could not, I must get one that could and would. I am not for making a man toil like a slave; but what would become of the world if a well-fed healthy man could exhaust himself in tilling and cropping and clearing half an acre of ground? I have known many men dig thirty rods of garden ground in a day; I have, before I was fourteen, digged twenty rods in a day, for more than ten days successively; and I have heard, and believe the fact, of a man at Portsea, who digged forty rods in one single day, between day-light and dark. So that it is no slavish toil that I am here recommending.

KEEPING PIGS

139. NEXT after the Cow comes the Pig; and, in many cases, where a cow cannot be kept, a pig or pigs may be kept. But these are animals not to be ventured on without due consideration as to the means of feeding them; for a starved pig is a great deal worse than none at all. You cannot make bacon as you can milk, merely out of the garden. There must be something more. A couple of flitches of bacon are worth fifty thousand Methodist sermons and religious tracts. The sight of them upon the rack tends more to keep a man from poaching and stealing than whole volumes of penal statutes, though assisted by the terrors of the hulks and gibbet. They are great softeners of the temper, and promoters of domestic harmony. They are a great blessing; but they are not to be had from a herbage or roots of

any kind; and therefore, before a pig be attempted, the means ought to be considered.

140. Breeding sows are great favourites with Cottagers in general; but I have seldom known them to answer their purpose. Where there is an outlet, the sow will, indeed, keep herself by grazing in summer, with a little wash to help her out; and when her pigs come, they are many in number; but they are a heavy expense. The sow must live as well as a fatting hog, or the pigs will be good for little. It is a great mistake, too, to suppose that the condition of the sow previous to pigging is of no consequence; and, indeed, some suppose that she ought to be rather bare of flesh at the pigging time. Never was a greater mistake; for if she be in this state, she presently becomes a mere rack of bones: and then, do what you will, the pigs will be poor things. However fat she may be before she farrow, the pigs will make her lean in a week. All her fat goes away in her milk, and unless the pigs have a store to draw upon they pull her down directly; and, by the time they are three weeks old, they are starving for want; and then they never come to good.

141. Now, a cottager's sow cannot, without great expense, be kept in a way to enable her to meet the demands of her farrow. She may look pretty well; but the flesh she has upon her is not of the same nature as that which the farm-yard sow carries about her. It is the result of grass, and of poor grass, too, or other weak food: and not made partly out of corn and whey and strong wash, as is the case of the farmer's sow. No food short of that of a fatting hog

will enable her to keep her pigs alive; and this she
must have for ten weeks, and that at a great expense.
Then comes the operation upon the principle of
Parson Malthus, in order to check population; and
there is some risk here, though not very great. But
there is the weaning; and who, that knows anything
about the matter, will think lightly of the weaning
of a farrow of pigs! By having nice food given them,
they seem, for a few days, not to miss their mother.
But their appearance soon shows the want of her.
Nothing but the very best food, and that given in
the most judicious manner, will keep them up to
any thing like good condition; and, indeed, there is
nothing short of milk that will effect the thing well.
How should it be otherwise! The very richest cow's
milk is poor compared with that of the sow; and, to
be taken from this and put upon food, one ingredient
of which is water, is quite sufficient to reduce the
poor little things to bare bones and staring hair, a
state to which cottagers' pigs very soon come in
general; and, at last, he frequently drives them to
market, and sells them for less than the cost of the
food which they and the sow have devoured since
they were farrowed. It was, doubtless, pigs of this
description that were sold the other day at Newbury
market, for fifteen pence a piece, and which were, I
dare say, dear even as a gift. To get such a pig to
begin to grow will require three months, and with
good feeding, too, in winter time. To be sure it does
come to be a hog at last; but, do what you can, it is
a dear hog.

142. The Cottager, then, can hold no competi-

tion with the Farmer in the breeding of pigs, to do which, with advantage, there must be milk, and milk, too, that can be advantageously applied to no other use. The cottager's pig must be bought ready weaned to his hand, and, indeed, at four months old, at which age, if he be in good condition, he will eat any thing that an old hog will eat. He will graze, eat cabbage leaves, and almost the stumps. Swedish turnip tops and roots, and such things, with a little wash, will keep him along in very good growing order. I have now to speak of the time of purchasing, the manner of keeping, of fatting, killing, and curing; but these I must reserve till my next Number.

No. VI

KEEPING PIGS

(*Continued*)

143. As in the case of cows so in that of pigs, much must depend upon the situation of the cottage; because all pigs will graze; and therefore, on the skirts of forests or commons, a couple or three pigs may be kept, if the family be considerable; and especially if the cottager brew his own beer, which will give him grains to assist the wash. Even in lanes, or on the sides of great roads, a pig will find a good part of his food from May to November; and if he be yoked, the occupiers of the neighbourhood must be churlish and brutish indeed, if they give the owner any annoyance.

144. Let me break off here for a moment to point out to my readers the truly excellent conduct of Lord WINCHELSEA and Lord STANHOPE, who, as I read, have taken great pains to make the labourers on their estates comfortable, by allotting to each a piece of ground sufficient for the keeping of a cow. I once, when I lived at Botley, proposed to the copyholders and other farmers in my neighbourhood, that we should petition the Bishop of Winchester, who was lord of the manors thereabouts, to grant titles to all the numerous persons called trespassers on the wastes; and also to give titles to others of the poor parishioners who were willing to make, on the skirts of the wastes, enclosures not exceeding an acre each. This I am convinced would have done a great deal towards relieving the parishes, then greatly burdened by men out of work. This would have been better than digging holes one day to fill them up the next. Not a single man would agree to my proposal! One, a bull-frog farmer (now, I hear, pretty well sweated down), said it would only make them saucy! And one, a true disciple of Malthus, said, that to facilitate their rearing children was a harm! This man had at the time, in his own occupation, land that had formerly been six farms, and he had, too, ten or a dozen children. I will not mention names; but this farmer will now, perhaps, have occasion to call to mind what I told him on that day, when his opposition, and particularly the ground of it, gave me the more pain, as he was a very industrious, civil, and honest man. Never was there a greater mistake than to suppose that men are made

saucy and idle by just and kind treatment. Slaves are always lazy and saucy; nothing but the lash will extort from them either labour or respectful deportment. I never met with a saucy Yankee (New Englander) in my life. Never servile, always civil. This must necessarily be the character of freemen living in a state of competence. They have nobody to envy, nobody to complain of, they are in good humour with mankind. It must, however, be confessed, that very little, comparatively speaking, is to be accomplished by the individual efforts even of benevolent men like the two noblemen before mentioned. They have a strife to maintain against the general tendency of the national state of things. It is by general and indirect means, and not by partial and direct and positive regulations, that so great a good as that which they generously aim at can be accomplished. When we are to see such means adopted God only knows; but if much longer delayed, I am of opinion that they will come too late to prevent something very much resembling a dissolution of society.

145. The cottager's pig should be bought in the spring, or late in winter; and being then four months old, he will be a year old before killing time; for it should always be borne in mind, that this age is required in order to ensure the greatest quantity of meat from a given quantity of food. If a hog be more than a year old, he is the better for it. The flesh is more solid and more nutritious than that of a young hog, much in the same degree that the mutton of a full-mouthed wether is better than that of a younger

wether. The pork or bacon of young hogs, even if fatted on corn, is very apt to boil out, as they call it; that is to say, come out of the pot smaller in bulk than it goes in. When you begin to fat, do it by degrees, especially in the case of hogs under a year old. If you feed high all at once, the hog is apt to surfeit, and then a great loss of food takes place. Peas, or barley-meal, is the food; the latter rather the best, and does the work quicker. Make him quite fat by all means. The last bushel, even if he sit as he eat, is the most profitable. If he can walk two hundred yards at a time, he is not well fatted. Lean bacon is the most wasteful thing that any family can use. In short, it is uneatable, except by drunkards, who want something to stimulate their sickly appetite. The man who cannot live on solid fat bacon, well fed and well cured, wants the sweet sauce of labour, or is fit for the hospital. But, then, it must be bacon, the effect of barley or peas (not beans), and not of whey, potatoes, or messes of any kind. It is frequently said, and I know that even farmers say it, that bacon made from corn costs more than it is worth! Why do they take care to have it then? They know better. They know well that it is the very cheapest they can have; and they who look at both ends and both sides of every cost would as soon think of shooting their hogs as of fatting them on messes; that is to say, for their own use, however willing they might now-and-then be to regale the Londoners with a bit of potato-pork.

146. About Christmas, if the weather be coldish, is a good time to kill. If the weather be very mild,

you may wait a little longer, for the hog cannot be too fat. The day before killing he should have no food. To kill a hog nicely is so much of a profession that it is better to pay a shilling for having it done, than to stab and hack and tear the carcass about. I shall not speak of pork, for I would by no means recommend it. They are two ways of going to work to make bacon; in the one you take off the hair by scalding. This is the practice in most parts of England, and all over America. But the Hampshire way, and the best way, is to burn the hair off. There is a great deal of difference in the consequences. The first method slackens the skin, opens all the pores of it, makes it loose and flabby by drawing out the roots of the hair. The second tightens the skin in every part, contracts all the sinews and veins in the skin, makes the flitch a solider thing, and the skin a better protection to the meat. The taste of the meat is very different from that of a scalded hog, and to this chiefly it was that Hampshire bacon owed its reputation for excellence. As the hair is to be burnt off it must be dry, and care must be taken that the hog be kept on dry litter of some sort the day previous to killing. When killed he is laid upon a narrow bed of straw, not wider than his carcass, and only two or three inches thick. He is then covered all over thinly with straw, to which, according as the wind may be, the fire is put at one end. As the straw burns, it burns the hair. It requires two or three coverings and burnings, and care is taken that the skin be not in any part burnt, or parched. When the hair is all burnt off close, the hog is scraped clean, but never touched with water. The upper side being finished the hog is turned over, and the other side is treated in like manner. This work should always be done before day-light, for in the day-light you cannot so nicely discover whether the hair be sufficiently burnt off. The light of the fire is weakened by that of the day. Besides, it makes the boys get up very early for once at any rate, and that is something, for boys always like a bonfire.

147. The inwards are next taken out, and if the wife be not a slattern, here, in the mere offal, in the mere garbage, there is food, and delicate food, too, for a large family for a week; and hog's puddings for the children, and some for neighbours' children, who come to play with them; for these things are by no means to be overlooked, seeing that they tend to the keeping alive of that affection in children for their parents, which later in life will be found absolutely necessary to give effect to wholesome precept, especially when opposed to the boisterous passions of youth.

148. The butcher the next day cuts the hog up, and then the house is filled with meat! Souse, griskins, blade-bones, thigh-bones, spare-ribs, chines, belly-pieces, cheeks, all coming into use one after the other, and the last of the latter not before the end of about four or five weeks. But about this time it is more than possible that the Methodist parson will pay you a visit. It is remarked in America, that these gentry are attracted by the squeaking of the pigs, as the fox is by the cackling of the hen. This may be called slander, but I will tell you what I did

know to happen. A good honest careful fellow had a spare-rib, on which he intended to sup with his family after a long and hard day's work at coppice-cutting. Home he came at dark with his two little boys, each with a nitch of wood that they had carried four miles, cheered with the thought of the repast that awaited them. In he went, found his wife, the Methodist parson, and a whole troop of the sisterhood, engaged in prayer, and on the table lay scattered the clean-polished bones of the spare-rib! Can any reasonable creature believe, that, to save the soul, God requires us to give up the food necessary to sustain the body? Did Saint Paul preach this? He who, while he spread the gospel abroad, worked himself, in order to have it to give to those who were unable to work? Upon what, then, do these modern saints, these evangelical gentlemen, found their claim to live on the labour of others?

149. All the other parts taken away, the two sides that remain, and that are called flitches, are to be cured for bacon. They are first rubbed with salt on their insides, or flesh sides, then placed one on the other, the flesh sides uppermost, in a salting trough which has a gutter round its edges to drain away the brine; for to have sweet and fine bacon the flitches must not lie sopping in brine, which gives it that sort of taste which barrel-pork and sea-jonk have, and than which nothing is more villainous. Every one knows how different is the taste of fresh dry salt from that of salt in a dissolved state. The one is savoury, the other nauseous. Therefore change the salt often. Once in four or five days. Let it melt and

sink in but let it not lie too long. Change the flitches. Put that at bottom which was first put on the top. Do this a couple of times. This mode will cost you a great deal more in salt, or rather in taxes, than the sopping mode; but without it your bacon will not be sweet and fine, and will not keep so well. As to the time required for making the flitches sufficiently salt, it depends on circumstances; the thickness of the flitch, the state of the weather, the place wherein the salting is going on. It takes a longer time for a thick than for a thin flitch; it takes longer in dry than in damp weather; it takes longer in a dry than a in damp place. But for the flitches of a hog of twelve score, in weather not very dry or very damp, about six weeks may do; and as yours is to be fat, which receives little injury from over-salting, give time enough; for you are to have bacon till Christmas comes again. The place for salting should, like a dairy, always be cool, but always admit of a free circulation of air; confined air, though cool, will taint meat sooner than the mid-day sun accompanied with a breeze. Ice will not melt in the hottest sun so soon as in a close and damp cellar. Put a lump of ice in cold water, and one of the same size before a hot fire, and the former will dissolve in half the time that the latter will. Let me take this occasion of observing, that an ice-house should never be under ground, or under the shade of trees. That the bed of it ought to be three feet above the level of the ground, that this bed ought to consist of something that will admit the drippings to go instantly off, and the house should stand in a place open to the sun and air. This is the way they have the ice-houses under the burning sun of Virginia; and here they keep their fish and meat as fresh and sweet as in winter, when at the same time neither will keep for twelve hours, though let down to the depth of a hundred feet in a well. A Virginian, with some poles and straw, will stick up an ice-house for ten dollars, worth a dozen of those ice-houses, each of which costs our men of taste as many scores of pounds. It is very hard to imagine, indeed, what any one should want ice for, in a country like this, except for clodpole boys to slide upon, and to drown cockneys in skiting-time; but if people must have ice in summer, they may as well go a right way as a wrong way to get it.

150. However, the patient that I have at this time under my hands wants nothing to cool his blood, but something to warm it, and therefore I will get back to the flitches of bacon, which are now to be smoked; for smoking is a great deal better than merely drying; as is the fashion in the dairy-countries in the West of England. When there were plenty of farm-houses, there were plenty of places to smoke bacon in; since farmers have lived in gentlemen's houses, and the main part of the farm-houses have been knocked down, these places are not so plenty. However, there is scarcely any neighbourhood without a chimney left to hang bacon up in. Two precautions are necessary: first, to hang the flitches where no rain comes down upon them: second, not to let them be so near the fire as to melt. These pre-cautions taken, the next is, that the smoke must

proceed from wood, not turf, peat, or coal. Stubble or litter might do: but the trouble would be great. Fir, or deal, smoke is not fit for the purpose. I take it, that the absence of wood, as fuel, in the dairy countries, and in the North, has led to the making of pork and dried bacon. As to the time that it requires to smoke a flitch, it must depend a good deal upon whether there be a constant fire beneath, and whether the fire be large or small. A month may do, if the fire be pretty constant, and such as a farm-house fire usually is. But over-smoking, or, rather, too long hanging in the air, makes the bacon rust. Great attention should, therefore, be paid to this matter. The flitch ought not to be dried up to the hardness of a board, and yet it ought to be perfectly dry. Before you hang it up, lay it on the floor, scatter the flesh side pretty thickly over with bran, or with some fine sawdust other than that of deal or fir. Rub it on the flesh, or pat it well down upon it. This keeps the smoke from getting into the little openings, and makes a sort of crust to be dried on; and, in short, keeps the flesh cleaner than it would otherwise be.

151. To keep the bacon sweet and good, and free from nasty things that they call hoppers; that is to say, a sort of skipping maggots, engendered by a fly which has a great relish for bacon; to provide against this mischief, and also to keep the bacon from becoming rusty, the Americans, whose country is so hot in summer, have two methods. They smoke no part of the hog except the hams, or gammons. They cover these with coarse linen cloth such as the finest

hop-bags are made of, which they sew neatly on. They then white-wash the cloth all over with lime white-wash, such as we put on walls, their lime being excellent stone lime. They give the ham four or five washings, the one succeeding as the former gets dry; and in the sun, all these washings are put on in a few hours. The flies cannot get through this; and thus the meat is preserved from them. The other mode, and that is the mode for you, is, to sift fine some clean and dry wood-ashes. Put some at the bottom of a box, or chest, which is long enough to hold a flitch of bacon. Lay in one flitch; then put in more ashes; then the other flitch; and then cover this with six or eight inches of the ashes. This will effectually keep away all flies; and will keep the bacon as fresh and good as when it came out of the chimney, which it will not be for any great length of time, if put on a rack, or kept hung up in the open air. Dust, or even sand, very, very dry, would, per-haps, do as well. The object is not only to keep out the flies, but the air. The place where the chest, or box, is kept, ought to be dry; and, if the ashes should get damp (as they are apt to do from the salts they contain), they should be put in the fire-place to dry, and then be put back again. Peat-ashes, or turf-ashes, might do very well for this purpose. With these precautions, the bacon will be as good at the end of the year as on the first day; and it will keep two, or even three years, perfectly good, for which, however, there can be no necessity.

152. Now, then, this hog is altogether a capital thing. The other parts will be meat for about four

or five weeks. The lard, nicely put down, will last a
long while for all the purposes for which it is wanted.
To make it keep well there should be some salt put
into it. Country children are badly brought up if
they do not like sweet lard spread upon bread, as we
spread butter. Many a score hunches of this sort
have I eaten, and I never knew what poverty was.
I have eaten it for luncheon, at the houses of good
substantial farmers in France and Flanders. I am
not now frequently so hungry as I ought to be; but
I should think it no hardship to eat sweet lard instead
of butter. But, now-a-days, the labourers, and
especially the female part of them, have fallen into
the taste of niceness in food and finery in dress; a
quarter of a bellyful and rags are the consequence.
The food of their choice is high-priced, so that, for
the greater part of their time, they are half-starved.
The dress of their choice is showy and flimsy, so
that, to-day, they are ladies, and to-morrow ragged
as sheep with the scab. But has not Nature made
the country girls as pretty as ladies? Oh, yes! (bless
their rosy cheeks and white teeth!) and a great deal
prettier, too! But are they less pretty, when their
dress is plain and substantial, and when the natural
presumption is, that they have smocks as well as
gowns, than they are when drawn off in the frail
fabric of Sir Robert Peel, " where tawdry colours
strive with dirty white," exciting violent sus-
picions that all is not as it ought to be nearer the
skin, and calling up a train of ideas extremely hostile
to that sort of feeling which every lass innocently
and commendably wishes to awaken in her male

beholders? Are they prettiest when they come through the wet and dirt safe and neat; or when their draggled dress is plastered to their backs by a shower of rain? However, the fault has not been theirs, nor that of their parents. It is the system of managing the affairs of the nation. This system has made all flashy and false, and has put all things out of their place. Pomposity, bombast, hyperbole, redundancy, and obscurity, both in speaking and in writing; mock-delicacy in manners, mock-liberality, mock-humanity, and mock-religion. Pitt's false money, Peel's flimsy dresses, Wilberforce's potato diet, Castlereagh's and Mackintosh's oratory, Walter Scott's poems, Walter's and Stoddart's paragraphs, with all the bad taste and baseness and hypocrisy which they spread over this country; all have arisen, grown, branched out, bloomed, and borne together; and we are now beginning to taste of their fruit. But, as the fat of the adder is, as is said, the antidote to its sting; so in the Son of the Great Worker of Spinning Jennies, we have, thanks to the Proctors and Doctors of Oxford, the author of that Bill, before which this false, this flashy, this flimsy, this rotten system will dissolve as one of his father's pasted calicoes does at the sight of the washing-tub!

153. "What," says the Cottager, "has all this to do with hogs and bacon?" Not directly with hogs and bacon, indeed; but it has a great deal to do, my good fellow, with your affairs, as I shall, probably, hereafter more fully show, though I shall now leave you to the enjoyment of your flitches of bacon, which, as I before observed, will do ten thousand

times more than any Methodist parson, or any other parson (except, of course, those of our church), to make you happy, not only in this world but in the world to come. Meat in the house is a great source of harmony, a great preventer of the temptation to commit those things which, from small beginnings, lead, finally, to the most fatal and atrocious results; and I hold that doctrine to be truly damnable which teaches that God has made any selection, any condition relative to belief, which is to save from punishment those who violate the principles of natural justice.

154. Some other meat you may have, but bacon is the great thing. It is always ready; as good cold as hot; goes to the field or the coppice conveniently; in harvest, and other busy times, demands the pot to be boiled only on a Sunday; has twice as much strength in it as any other thing of the same weight; and, in short, has in it every quality that tends to make a labourer's family able to work and be well off. One pound of bacon, such as that which I have described, is, in a labourer's family, worth four of five of ordinary mutton or beef, which are great part bone, and which, in short, are gone in a moment. But always observe, it is fat bacon that I am talking about. There will, in spite of all that can be done, be some lean in the gammons, though comparatively very little, and therefore you ought to begin at that end of the flitches, for old lean bacon is not good.

155. Now, as to the cost. A pig (a spayed sow is best) bought in March, four months old, can be had

now for fifteen shillings. The cost till fatting time is
next to nothing to a Cottager; and then the cost, at
the present price of corn, would, for a hog of twelve
score, not exceed three pounds; in the whole four
pounds five; a pot of poison a week bought at the
public-house comes to twenty-six shillings of the
money; and more than three times the remainder
is generally flung away upon the miserable tea, as
I have clearly shown in the first Number, at paragraph
24. I have, indeed, there shown, that if the tea were
laid aside, the labourer might supply his family well
with beer all the year round, and have a fat hog of
even fifteen score for the cost of the tea, which does
him, and can do him, no good at all.

156. The feet, the cheeks, and other bone, being
considered, the bacon and lard, taken together,
would not exceed sixpence a pound. Irish bacon is
" cheaper." Yes, lower-priced. But I will engage
that a pound of mine, when it comes out of the pot
(to say nothing of the taste), shall weigh as much as
a pound and a half of Irish, or any dairy or slop-fed
bacon, when that comes out of the pot. No, no: the
farmers joke when they say that their bacon costs
them more than they could buy bacon for. They
know well what it is they are doing; and besides,
they always forget, or rather, remember not to say,
that the fatting of a large hog yields them three or
four load of dung, really worth more than ten or
fifteen of common yard dung. In short, without hogs,
farming could not go on; and it never has gone on
in any country in the world. The hogs are the great
stay of the whole concern. They are much in small

space; they make no show, as flocks and herds do; but without them the cultivation of the land would be a poor, a miserably barren concern.

SALTING MUTTON AND BEEF

157. VERY FAT Mutton may be salted to great advantage, and also smoked, and may be kept thus a long while. Not the shoulders and legs, but the back of the sheep. I have never made any flitch of sheep-bacon, but I will, for there is nothing like having a store of meat in a house. The running to the butcher's daily is a ridiculous thing. The very idea of being fed, of a family, being fed, by daily supplies, has something in it perfectly tormenting. One half of the time of a mistress of a house, the affairs of which are carried on in this way, is taken up in talking about what is to be got for dinner, and in negotiations with the butcher. One single moment spent at table beyond what is absolutely necessary is a moment very shamefully spent; but to suffer a system of domestic economy, which unnecessarily wastes daily an hour or two of the mistress's time in hunting for the provision for the repast, is a shame indeed; and when we consider how much time is generally spent in this and in equally absurd ways, it is no wonder that we see so little performed by numerous individuals as they do perform during the course of their lives.

158. Very fat parts of Beef may be salted and smoked in a like manner. Not the lean, for that is a

great waste, and is, in short, good for nothing. Poor fellows on board of ships are compelled to eat it, but it is a very bad thing.

<hr>

No. VII

BEES, FOWLS, &c., &c.

159. I NOW proceed to treat of objects of less importance than the foregoing, but still such as may be worthy of great attention. If all of them cannot be expected to come within the scope of a labourer's family, some of them must and others may; and it is always of great consequence that children be brought up to set a just value upon all useful things, and especially upon all living things; to know the utility of them: for without this they never, when grown up, are worthy of being entrusted with the care of them. One of the greatest, and perhaps the very commonest, faults of servants, is, their inadequate care of animals committed to their charge. It is a well-known saying that " the master's eye makes the horse fat," and the remissness to which this alludes is generally owing to the servant not having been brought up to feel an interest in the well-being of animals.

<hr>

BEES

160. IT is not my intention to enter into a history of this insect, about which so much has been written, especially by the French naturalists. It is the useful

that I shall treat of, and that is done in not many words. The best hives are those made of clean un-blighted rye-straw. Boards are too cold in England. A swarm should always be put in a new hive, and the sticks should be new that are put into the hive for the bees to work on; for, if the hive be old, it is not so wholesome, and a thousand to one but it contain the embryos of moths and other insects injurious to bees. Over the hive itself there should be a cap of thatch made also of clean rye-straw; and it should not only be new when first put on the hive, but a new one should be made to supply the place of the former one every three or four months; for when the straw begins to get rotten, as it soon does, insects breed in it, its smell is bad, and its effect on the bees is dangerous.

161. The hives should be placed on a bench, the legs of which mice and rats cannot creep up. Tin round the legs is best. But even this will not keep down ants, which are mortal enemies of bees. To keep these away, if you find them infest the hive, take a green stick and twist it round in the shape of a ring to lie on the ground round the leg of the bench and at a few inches from it, and cover this stick with tar. This will keep away the ants. If the ants come from one home, you may easily trace them to it; and when you have found it, pour boiling water on it in the night, when all the family are at home. This is the only effectual way of destroying ants, which are frequently so troublesome. It would be cruel to cause this destruction, if it were not necessary to do it, in order to preserve the honey, and indeed the bees too.

162. Besides the hive and its cap, there should be a sort of shed, with top, back, and ends, to give additional protection in winter; though in summer hives may be kept too hot, and in that case the bees become sickly and the produce becomes light. The situation of the hive is to face the South-east; or at any rate, to be sheltered from the North and the West. From the North always, and from the West in winter. If it be a very dry season in summer, it contributes greatly to the success of the bees to place clear water near their home, in a thing that they can conveniently drink out of; for if they have to go a great way for drink, they have not much time for work.

163. It is supposed that bees live only a year; at any rate it is best never to keep the same stall, or family, over two years, except you want to increase your number of hives. The swarm of this summer should always be taken in the autumn of next year. It is whimsical to save the bees when you take the honey. You must feed them; and if saved, they will die of old age before the next fall; and though young ones will supply the place of the dead, this is nothing like a good swarm put up during the summer.

164. As to the things that bees make their collections from, we do not perhaps know a thousandth part of them; but of all the blossoms that they seek eagerly, that of the Buck-wheat stands foremost. Go round a piece of this grain just towards sunset, when the buck-wheat is in bloom, and you will see the air filled with bees going home from it in all directions. The buck-wheat, too, continues in bloom a long while, for the grain is dead ripe on one part

of the plant, while there are fresh blossoms coming
out on the other part.

165. A good stall of bees, that is to say, the produce of one, is always worth about two bushels of
good wheat. The cost is nothing to the labourer.
He must be a stupid countryman indeed who cannot
make a bee-hive; and a lazy one indeed if he will not,
if he can. In short, there is nothing but care demanded; and there are very few situations in the
country, especially in the south of England, where a
labouring man may not have half a dozen stalls of
bees to take every year. The main things are to keep
away insects, mice, and birds, and especially a little
bird called the bee-bird; and to keep all clean and
fresh as to the hives and coverings. Never put a
swarm into an old hive. If wasps, or hornets, annoy
you, watch them home in the daytime; and in the
night kill them by fire, or by boiling water. Fowls
should not go where bees are, for they will eat them.

166. Suppose a man may get three stalls of bees
in a year. Six bushels of wheat give him bread for
an eighth part of the year. Scarcely anything is a
greater misfortune than shiftlessness. It is an evil
little short of the loss of eyes or of limbs.

GEESE

167. THEY can be kept to advantage only where
there are green commons, and there they are easily
kept, live to a very great age, and are amongst the
hardiest animals in the world. If kept well, a goose

will lay a hundred eggs in a year. The French put their eggs under large hens of common fowls, to each of which they give four or five eggs; or under turkeys, to which they give nine or ten goose-eggs. If the goose herself sit, she must be well and regularly fed at, or near to, her nest. When the young ones are hatched, they should be kept in a warm place for about four days, and fed on barley-meal, mixed, if possible, with milk; and then they will begin to graze. Water for them, or for the old ones to swim in, is by no means necessary, nor perhaps, ever even useful. Or how is it that you see such fine flocks of fine geese all over Long Island (in America), where there is scarcely such a thing as a pond or a run of water?

168. Geese are raised by grazing, but to fat them something more is required. Corn of some sort, or boiled Swedish turnips. Some corn and some raw Swedish turnips, or carrots, or white cabbage, or lettuces, make the best fatting. The modes that are resorted to by the French for fatting geese, nailing them down by their webs, and other acts of cruelty, are, I hope, such as Englishmen never think of. They will get fat enough without the use of any of these unfeeling means being employed. He who can deliberately inflict torture upon an animal, in order to heighten the pleasure his palate is to receive in eating it, is an abuser of the authority which God has given him, and is, indeed, a tyrant in his heart. Who would think himself safe, if at the mercy of such a man? Since the first edition of this work was published I have had a good deal of experience

with regard to geese. It is a very great error to suppose that what is called a Michaelmas goose is the
thing. Geese are, in general, eaten at the age when
they are called green geese; or after they have got
their full and entire growth, which is not until the
latter part of October. Green geese are tasteless
squabs; loose flabby things; no rich taste in them;
and, in short, a very indifferent sort of dish. The
full grown goose has solidity in it; but it is hard, as
well as solid; and in place of being rich, it is strong.
Now, there is a middle course to take; and if you
take this course, you produce the finest birds of
which we can know anything in England. For three
years, including the present year, I have had the
finest geese that I ever saw, or ever heard of. I have
bought from twenty to thirty every one of these
years. I buy them off the common late in June, or
very early in July. They have cost me from two
shillings to three shillings each, first purchase. I
bring the flock home, and put them in a pen, about
twenty feet square, where I keep them well littered
with straw, so as for them not to get filthy. They
have one trough in which I give them dry oats, and
they have another trough where they have constantly
plenty of clean water. Besides these, we give them,
two or three times a day, a parcel of lettuces out of
the garden. We give them such as are going to seed
generally; but the better the lettuces are, the better
the geese. If we have no lettuces to spare, we give
them cabbages, either loaved or not loaved; though,
observe, the white cabbage as well as the white
lettuce, that is to say, the loaved cabbage and lettuce,

are a great deal better than those that are not loaved. This is the food of my geese. They thrive exceedingly upon this food. After we have had the flock about ten days, we begin to kill, and we proceed once or twice a week till about the middle of October, sometimes later. A great number of persons who have eaten of these geese have all declared that they did not imagine that a goose could be brought to be so good a bird. These geese are altogether different from the hard, strong things that come out of the stubble fields, and equally different from the flabby things called a green goose. I should think that the cabbages or lettuces perform half the work of keeping and fatting my geese; and these are things that really cost nothing. I should think that the geese, upon an average, do not consume more than a shilling's worth of oats each. So that we have these beautiful geese for about four shillings each. No money will buy me such a goose in London; but the thing that I can get nearest to it will cost me seven shillings. Every gentleman has a garden. That garden has, in the month of July, a wagon-load, at least, of lettuces and cabbages to throw away. Nothing is attended with so little trouble as these geese. There is hardly anybody near London that has not room for the purposes here mentioned. The reader will be apt to exclaim, as my friends very often do, " Cobbett's Geese are all Swans." Well, better that way than not to be pleased with what one has. However, let gentlemen try this method of fatting geese. It saves money, mind, at the same time. Let them try it; and if any one, who shall try it, shall find the

effect not to be that which I say it is, let him reproach
me publicly with being a deceiver. The thing is no
invention of mine. While I could buy a goose off
the common for half-a-crown, I did not like to give
seven shillings for one in London, and yet I wished
that geese should not be excluded from my house.
Therefore I bought a flock of geese, and brought
them home to Kensington. They could not be eaten
all at once. It was necessary, therefore, to fix upon
a mode of feeding them. The above mode was
adopted by my servant, as far as I know, without
any knowledge of mine; but the very agreeable result
made me look into the matter; and my opinion, that
the information will be useful to many persons, at
any rate, is sufficient to induce me to communicate
it to my readers.

DUCKS

169. No water to swim in, is necessary to the old,
and is injurious to the very young. They never should
be suffered to swim (if water be near) till more than
a month old. The old duck will lay, in the year, if
well kept, ten dozen of eggs; and that is her best
employment; for common hens are the best mothers.
It is not good to let young duck out in the morning
to eat slugs and worms, for though they like them,
these things kill them if they eat a great quantity.
Grass, corn, white cabbages, and lettuces, and
especially buck-wheat, cut, when half ripe, and

flung down in the haulm. This makes fine ducks. Ducks will feed on garbage and all sorts of filthy things; but their flesh is strong, and bad in proportion. They are, in Long Island, fatted upon a coarse sort of crab called a horse-foot fish, prodigious quantities of which are cast on the shores. The young ducks grow very fast upon this, and very fat; but woe unto him that has to smell them when they come from the spit; and, as for eating them, a man must have a stomach indeed to do that!

170. When young, they should be fed upon barley-meal, or curds, and kept in a warm place in the night-time, and not let out early in the morning. They should, if possible, be kept from water to swim in. It always does them harm, and, if intended to be sold to be killed young, they should never go near ponds, ditches, or streams. When you come to fat ducks, you must take care that they get at no filth whatever. They will eat garbage of all sorts; they will suck down the most nauseous particles of all those substances which go for manure. A dead rat three parts rotten is a feast to them. For these reasons I should never eat any ducks, unless there were some mode of keeping them from this horrible food. I treat them precisely as I do my geese. I buy a troop when they are young, and put them in a pen, and feed them upon oats, cabbages, lettuces, and water, and have the place kept very clean. My ducks are, in consequence of this, a great deal more fine and delicate than any others that I know anything of.

TURKEYS

171. THESE are flying things, and so are common fowls. But it may happen that a few hints respecting them may be of use. To raise turkeys, in this chilly climate, is a matter of much greater difficulty than in the climates that give great warmth. But the great enemy to young turkeys (for old ones are hardy enough) is the wet. This they will endure in no climate; and so true is this, that, in America, where there is always " a wet spell," in April, the farmers' wives take care never to have a brood come out until the spell is passed. In England, where the wet spells come at haphazard, the first thing is to take care that young turkeys never go out, on any account, except in dry weather, till the dew be quite off the ground; and this should be adhered to till they get to be of the size of an old partridge, and have their backs well covered with feathers. And, in wet weather, they should be kept under cover all day long.

172. As to the feeding of them, when young, various nice things are recommended. Hard eggs chopped fine, with crumbs of bread, and a great many other things; but that which I have seen used, and always with success, and for all sorts of young poultry, is milk turned into curds. This is the food for young poultry of all sorts. Some should be made fresh every day; and if this be done, and the young turkeys kept warm, and especially from wet, not one out of a score will die. When they get to be strong, they may have meal and grain, but still they always love the curds.

173. When they get their head feathers they are hardy enough; and what they then want is room to prowl about. It is best to breed them under a common hen; because she does not ramble like a hen-turkey; and it is a very curious thing that the turkeys bred up by a hen of the common fowl, do not themselves ramble much when they get old; and for this reason, when they buy turkeys for stock, in America (where there are such large woods, and where the distant rambling of turkeys is inconvenient), they always buy such as have been bred under the hens of the common fowl; than which a more complete proof of the great powers of habit is, perhaps, not to be found. And ought not this to be a lesson to fathers and mothers of families? Ought not they to consider that the habits which they give their children are to stick by those children during their whole lives?

174. The hen should be fed exceedingly well, too, while she is sitting and after she has hatched; for though she does not give milk, she gives heat; and, let it be observed, that as no man ever yet saw healthy pigs with a poor sow, so no man ever saw healthy chickens with a poor hen. This is a matter much too little thought of in the rearing of poultry; but it is a matter of the greatest consequence. Never let a poor hen sit; feed the hen well while she is sitting, and feed her most abundantly when she has young ones; for then her labour is very great; she is making exertions of some sort or other during the whole twenty-four hours; she has no rest; is constantly doing something or other to provide food or safety for her young ones.

175. As to fatting turkeys, the best way is, never to let them be poor. Cramming is a nasty thing, and quite unnecessary. Barley-meal, mixed with skim-milk, given to them, fresh and fresh, will make them fat in a short time, either in a coop, in a house, or running about. Boiled carrots and Swedish turnips will help, and it is a change of sweet food. In France they sometimes pick turkeys alive, to make them tender; of which I shall only say, that the man that can do this, or order it to be done, ought to be skinned alive himself.

FOWLS

176. THESE are kept for two objects; their flesh and their eggs. As to rearing them, everything said about rearing turkeys is applicable here. They are best fatted, too, in the same manner. But, as to lay-ing-hens, there are some means to be used to secure the use of them in winter. They ought not to be old hens. Pullets, that is, birds hatched in the foregoing spring, are, perhaps, the best. At any rate, let them not be more than two years old. They should be kept in a warm place, and not let out, even in the day time, in wet weather; for one good sound wetting will keep them back for a fortnight. The dry cold, even in the severest cold, if dry, is less injurious than even a little wet in winter time. If the feathers get wet, in our climate, in winter, or in short days, they do not get dry for a long time; and this it is that spoils and kills many of our fowls.

177. The French, who are great egg eaters, take singular pains as to the food of laying-hens in winter. They let them out very little, even in their fine climate, and give them very stimulating food; barley boiled, and given them warm; curds, buck-wheat (which, I believe, is the best thing of all except curds); parsley and other herbs chopped fine; leeks chopped in the same way; also apples and pears chopped very fine; oats and wheat cribbled; and sometimes they give them hemp-seed, and the seed of nettles; or dried nettles harvested in summer, and boiled in the winter. Some give them ordinary food, and, once a day, toasted bread sopped in wine. White cabbages chopped up are very good in winter for all sorts of poultry.

178. This is taking a great deal of pains; but the produce is also great and very valuable in winter; for, as to preserved eggs, they are things to run from and not after. All this supposes, however, a proper hen-house, about which we, in England, take very little pains. The vermin, that is to say, the lice, that poultry breed, are the greatest annoyance. And as our wet climate furnishes them, for a great part of the year, with no dust by which to get rid of these vermin, we should be very careful about cleanliness in the hen-houses. Many a hen, when sitting, is compelled to quit her nest to get rid of the lice. They torment the young chickens. And, in short, are a great injury. The fowl-house should, therefore, be very often cleaned out; and sand, or fresh earth, should be thrown on the floor. The nest should not be on shelves, or anything fixed; but little flat baskets

something like those that the gardeners have in the markets in London, and which they call sieves, should should be placed against the sides of the house upon pieces of wood nailed up for the purpose. By this means the nests are kept perfectly clean, because the baskets are, when necessary, taken down, the hay thrown out, and the baskets washed; which cannot be done, if the nest be made in anything forming a part of the building. Besides this the roosts ought to be cleaned every week, and the hay changed in the nests of laying-hens. It is good to fumigate the house frequently by burning dry herbs, juniper wood, cedar wood, or with brimstone; for nothing stands so much in need of cleanliness as a fowl-house, in order to have fine fowls and plenty of eggs.

179. The ailments of fowls are numerous, but they would seldom be seen, if the proper care were taken. It is useless to talk of remedies in a case where you have complete power to prevent the evil. If well fed, and left perfectly clean, fowls will seldom be sick; and, as to old age, they never ought to be kept more than a couple or three years; for they get to be good for little as layers, and no teeth can face them as food.

180. It is, perhaps, seldom that fowls can be kept conveniently about a cottage; but when they can, three, four, or half a dozen hens to lay in winter, when the wife is at home the greater part of the time, are worth attention. They would require but little room, might be bought in November and sold in April, and six of them, with proper care, might be made to clear every week the price of a gallon of

flour. If the labour were great, I should not think
of it; but it is none; and I am for neglecting nothing
in the way of pains in order to ensure a hot dinner
every day in winter, when the man comes home
from work. As to the fatting of fowls, information
can be of no use to those who live in a cottage all
their lives; but it may be of some use to those who
are born in cottages and go to have the care of poultry
at richer persons' houses. Fowls should be put to
fat about a fortnight before they are wanted to be
killed. The best food is barley-meal wetted with
milk, but not wetted too much. They should have
clear water to drink, and it should be frequently
changed. Crammed fowls are very nasty things:
but " barn-door " fowls, as they are called, are
sometimes a great deal more nasty. *Barn*-door would,
indeed, do exceedingly well; but it unfortunately
happens that the stable is generally pretty near to
the barn. And now let any gentleman who talks
about sweet barn-door fowls, have one caught in
the yard, where the stable is also. Let him have it
brought in, killed, and the craw taken out and cut
open. Then let him take a ball of horse-dung from
the stable-door; and let his nose tell him how very
small is the difference between the smell of the
horse-dung and the smell of the craw of his fowl.
In short, roast the fowl, and then pull aside the skin
at the neck, put your nose to the place, and you will
almost think you are at the stable-door. Hence the
necessity of taking them away from the barn-door
a fortnight, at least, before they are killed. We know
very well that ducks that have been fed upon fish,

either wild ducks or tame ducks, will scent a whole room, and drive out of it all those who have not got pretty good constitutions. It must be so. Solomon says that all flesh is grass; and those who know anything about beef, know the difference between the effect of the grass in Herefordshire and Lincolnshire, and the effect of turnips and oil cake. In America they always take the fowls from the farmyard, and shut them up a fortnight or three weeks before they be killed. One thing, however, about fowls ought always to be borne in mind. They are never good for anything when they have attained their full growth, unless they be capons or poullards. If the poulets be old enough to have little eggs in them, they are not worth one farthing; and as to the cocks of the same age, they are fit for nothing but to make soup for soldiers on their march, and they ought to be taken for that purpose.

PIGEONS

181. A FEW of these may be kept about any cottage, for they are kept even in towns by labourers and artizans. They cause but little trouble. They take care of their own young ones; and they do not scratch, or do any other mischief in gardens. They want feeding with tares, peas, or small beans; and buck-wheat is very good for them. To begin keeping them, they must not have flown at large before you get them. You must keep them for two or three days, shut into the place which is to be their home;

and then they may be let out, and will never leave
you, as long as they can get proper food, and are
undisturbed by vermin, or unannoyed exceedingly
by lice.

182. The common dove-house pigeons are the
best to keep. They breed oftenest, and feed their
young ones best. They begin to breed at about nine
months old, and if well kept, they will give you eight
or nine pair in the year. Any little place, a shelf in
the cow shed; a board or two under the eaves of the
house; or, in short, any place under cover, even on
the ground floor, they will sit and hatch and breed
up their young ones in.

183. It is not supposed that there could be much
profit attached to them; but they are of this use;
they are very pretty creatures; very interesting in
their manners; they are an object to delight children,
and to give them the early habit of fondness for
animals and of setting a value on them, which, as I
have often had to observe before, is a very great
thing. A considerable part of all the property of a
nation consists of animals. Of course a proportionate
part of the cares and labours of a people appertains
to the breeding and bringing to perfection those
animals; and, if you consult your experience, you
will find that a labourer is, generally speaking, of
value in proportion as he is worthy of being intrusted
with the care of animals. The most careless fellow
cannot hurt a hedge or ditch; but to trust him with
the team, or the flock, is another matter. And, mind,
for the man to be trustworthy in this respect, the
boy must have been in the habit of being kind and

considerate towards animals; and nothing is so likely to give him that excellent habit as his seeing, from his very birth, animals taken great care of, and treated with great kindness, by his parents, and now-and-then having a little thing to call his own.

RABBITS

184. IN this case, too, the chief use, perhaps, is to give children those habits of which I have been just speaking. Nevertheless, rabbits are really profitable. Three does and a buck will give you a rabbit to eat for every three days in the year, which is a much larger quantity of food than any man will get by spending half his time in the pursuit of wild animals, to say nothing of the toil, the tearing of clothes, and the danger of pursuing the latter.

185. Everybody knows how to knock up a rabbit hutch. The does should not be allowed to have more than seven litters in a year. Six young ones to a doe is all that ought to be kept; and then they will be fine. Abundant food is the main thing; and what is there that a rabbit will not eat? I know of nothing green that they will not eat; and if hard pushed, they will eat bark, and even wood. The best thing to feed the young ones on when taken from the mother, is the carrot, wild or garden, parsnips, Swedish turnips, roots of dandelion; for too much green or watery stuff is not good for weaning rabbits. They should remain as long as possible with the mother. They should have oats once a day; and,

after a time, they may eat anything with safety. But if you give them too much green at first when they are weaned, they rot as sheep do. A variety of food is a great thing; and, surely, the fields and gardens and hedges furnish this variety! All sorts of grasses, strawberry-leaves, ivy, dandelions, the hog-weed or wild parsnip, in root, stem, and leaves. I have fed working horses, six or eight in number, upon this plant for weeks together. It is a tall bold plant, that grows in prodigious quantities in the hedges and coppices in some parts of England. It is the perennial parsnip. It has flower and seed precisely like those of the parsnip; and hogs, cows, and horses, are equally fond of it. Many a half-starved pig have I seen within a few yards of cart-loads of this pig-meat! This arises from want of the early habit of attention to such matters. I, who used to get hog-weed for pigs and for rabbits when a little chap, have never forgotten that the wild parsnip is good for pigs and rabbits.

186. When the doe has young ones, feed her most abundantly with all sorts of greens and herbage, and with carrots and the other things mentioned before, besides giving her a few oats once a day. That is the way to have fine healthy young ones, which, if they come from the mother in good case, will very seldom die. But do not think that because she is a small animal, a little feeding is sufficient. Rabbits eat a great deal more than cows or sheep, in proportion to their bulk.

187. Of all animals rabbits are those that boys are most fond of. They are extremely pretty, nimble in

their movements, engaging in their attitudes, and always completely under immediate control. The produce has not long to be waited for. In short, they keep an interest constantly alive in a little chap's mind; and they really cost nothing; for as to the oats, where is the boy that cannot, in harvest-time, pick up enough along the lanes to serve his rabbits for a year? The care is all; and the habit of taking care of things is, of itself, a most valuable possession.

188. To those gentlemen who keep rabbits for the use of their family (and a very useful and convenient article they are), I would observe, that when they find their rabbits die, they may depend on it, that ninety-nine times out of the hundred starvation is the malady. And particularly short feeding of the doe, while and before she has young ones; that is to say, short feeding of her at all times; for, if she be poor, the young ones will be good for nothing. She will live being poor, but she will not and cannot breed up fine young ones.

GOATS AND EWES

189. IN some places where a cow cannot be kept, a goat may. A correspondent points out to me, that a Dorset ewe or two might be kept on a common near a cottage to give milk; and certainly this might be done very well; but I should prefer a goat, which is hardier and much more domestic. When I was in the army, in New Brunswick, where, be it observed, the snow lies on the ground seven months in the

year, there were many goats that belonged to the regiment, and that went about with it on shipboard and everywhere else. Some of them had gone through nearly the whole of the American War. We never fed them. In summer they picked about wherever they could find grass; and in winter they lived upon cabbage-leaves, turnip-peelings, potato-peelings, and other things flung out of the soldiers' rooms and huts. One of these goats belonged to me, and, on an average throughout the year, she gave me more than three half-pints of milk a day. I used to have the kid killed when a few days old; and, for some time, the goat would give, nearly or quite, two quarts of milk a day. She was seldom dry more than three weeks in the year.

190. There is one great inconvenience belonging to goats; that is, they bark all young trees that they come near; so that, if they get into a garden, they destroy everything. But there are seldom trees on commons, except such as are too large to be injured by goats; and I can see no reason against keeping a goat where a cow cannot be kept. Nothing is so hardy; nothing is so little nice as to its food. Goats will pick peelings out of the kennel and eat them. They will eat mouldy bread or biscuit; fusty hay, and almost rotten straw; furze-bushes, heath thistles, and, indeed, what will they not eat, when they will make a hearty meal on paper, brown or white, printed on or not printed on, and give milk all the while! They will lie in any dog-hole. They do very well clogged, or stumped out. And then, they are very healthy things into the bargain, however closely they may

be confined. When sea voyages are so stormy as to kill geese, ducks, fowls, and almost pigs, the goats are well and lively; and when a dog of no kind can keep the deck for a minute, a goat will skip about upon it as bold as brass.

191. Goats do not ramble from home. They come in regularly in the evening, and if called, they come, like dogs. Now, though ewes, when taken great care of, will be very gentle, and though their milk may be rather more delicate than that of the goat, the ewes must be fed with nice and clean food, and they will not do much in the milk-giving way upon a common; and, as to feeding them, provision must be made pretty nearly as for a cow. They will not endure confinement like goats; and they are subject to numerous ailments that goats know nothing of. Then the ewes are done by the time they are about six years old; for they then lose their teeth; whereas the goat will continue to breed and to give milk in abundance for a great many years. The sheep is frightened at everything, and especially at the least sound of a dog. A goat, on the contrary, will face a dog, and, if he be not a big and courageous one, beat him off.

192. I have often wondered how it happened that none of our labourers kept goats; and I really should be glad to see the thing tried. They are pretty creatures, domestic as a dog, will stand and watch, as a dog does, for a crumb of bread, as you are eating; give you no trouble in the milking; and I cannot help being of opinion, that it might be of great use to introduce them amongst our labourers.

CANDLES AND RUSHES

193. WE are not permitted to make candles ourselves, and if we were, they ought seldom to be used
in a labourer's family. I was bred and brought up
mostly by rush-light, and I do not find that I see less
clearly than other people. Candles certainly were
not much used in English labourers' dwellings in
the days when they had meat dinners and Sunday
coats. Potatoes and taxed candles seem to have grown
into fashion together; and, perhaps, for this reason:
that when the pot ceased to afford grease for the
rushes, the potato-gorger was compelled to go to
the chandler's shop for light to swallow the potatoes
by, else he might have devoured peeling and all!

194. My grandmother, who lived to be pretty
nearly ninety, never, I believe, burnt a candle in
her house in her life. I know that I never saw one
there, and she, in a great measure, brought me up.
She used to get the meadow-rushes, such as they
tie the hop-shoots to the poles with. She cut them
when they had attained their full substance, but were
still green. The rush, at this age, consists of a body
of pith with a green skin on it. You cut off both ends
of the rush, and leave the prime part, which, on an
average, may be about a foot and a half long. Then
you take off all the green skin, except for about a
fifth part of the way round the pith. Thus it is a piece
of pith, all but a little strip of skin in one part, all
the way up, which, observe, is necessary to hold the
pith together all the way along.

195. The rushes being thus prepared, the grease

is melted, and put in a melted state into something
that is as long as the rushes are. The rushes are put
into the grease; soaked in it sufficiently; then taken
out and laid in a bit of bark taken from a young tree,
so as not to be too large. This bark is fixed up
against the wall by a couple of straps put round it;
and there it hangs for the purpose of holding the
rushes.

196. The rushes are carried about in the hand;
but to sit by, to work by, or to go to bed by, they
are fixed in stands made for the purpose, some of
which are high, to stand on the ground, and some
low, to stand on a table. The stands have an iron
port something like a pair of pliers to hold the rush
in, and the rush is shifted forward from time to time,
as it burns down to the thing that holds it.

197. Now these rushes give a better light than a
common small dip-candle; and they cost next to
nothing, though the labourer may with them have
as much light as he pleases, and though without them
he must sit the far greater part of the winter evenings
in the dark, even if he expend fifteen shillings a year
in candles. You may do any sort of work by this
light; and, if reading be your taste, you may read
the foul libels, the lies and abuse, which are circu-
lated gratis about me by the " Society for Promoting
Christian Knowledge," as well by rush-light as you
can by the light of taxed candles; and at any rate,
you would have one less evil; for to be deceived and
to pay a tax for the deception are a little too much
for even modern loyalty openly to demand.

MUSTARD

198. Why buy this, when you can grow it in your garden? The stuff you buy is half drugs, and is injurious to health. A yard square of ground, sown with common Mustard, the crop of which you would grind for use, in a little mustard-mill, as you wanted it, would save you some money, and probably save your life. Your mustard would look brown instead of yellow; but the former colour is as good as the latter; and, as to the taste, the real mustard has certainly a much better than that of the drugs and flour which go under the name of mustard. Let any one try it, and I am sure he will never use the drugs again. The drugs, if you take them freely, leave a burning at the pit of your stomach, which the real mustard does not.

DRESS, HOUSEHOLD GOODS, AND FUEL

199. In paragraph 152, I said, I think, enough to caution you, the English labourer, against the taste, now too prevalent, for fine and flimsy dress. It was, for hundreds of years, amongst the characteristics of the English people, that their taste was, in all matters, for things solid, sound, and good; for the useful, and decent, the cleanly in dress, and not for the showy. Let us hope that this may be the taste again; and let us, my friends, fear no troubles, no perils, that may be necessary to produce a return of that taste, accompanied with full bellies and warm backs to the labouring classes.

200. In household goods, the warm, the strong, the durable, ought always to be kept in view. Oak tables, bedsteads and stools, chairs of oak or of yew-tree, and never a bit of miserable deal board. Things of this sort ought to last several lifetimes. A labourer ought to inherit from his great-grandfather something beside his toil. As to bedding, and other things of that sort, all ought to be good in their nature, of a durable quality, and plain in their colour and form. The plates, dishes, mugs, and things of that kind, should be of pewter, or even of wood. Anything is better than crockery-ware. Bottles to carry a-field should be of wood. Formerly, nobody but the gipsies and mumpers, that went a hop-picking in the season, carried glass or earthen bottles. As to glass of any sort, I do not know what business it has in any man's house, unless he be rich enough to live on his means. It pays a tax, in many cases, to the amount of two-thirds of its cost. In short, when a house is once furnished with sufficient goods, there ought to be no renewal of hardly any part of them wanted for half an age, except in case of destruction by fire. Good management in this way leaves the man's wages to provide an abundance of good food and good raiment; and these are the things that make happy families; these are the things that make a good, kind, sincere, and brave people; not little pamphlets about " loyalty " and " content." A good man will be contented fast enough, if he be fed and clad sufficiently; but if a man be not well fed and clad, he is a base wretch to be contented.

201. Fuel should be, if possible, provided in

summer, or at least some of it. Turf and peat must be got in summer, and some wood may. In the woodland countries, the next winter ought to be thought of in June, when people hardly know what to do with the fuelwood; and something should, if possible, be saved in the bark-harvest to get a part of the fuel for the next winter. Fire is a capital article. To have no fire, or a bad fire, to sit by, is a most dismal thing. In such a state man and wife must be something out of the common way to be in good humour with each other, to say nothing of colds and other ailments which are the natural consequence of such misery. If we suppose the great Creator to condescend to survey his works in detail, what object can be so pleasing to him as that of the labourer, after his return from the toils of a cold winter day, sitting with his wife and children round a cheerful fire, while the wind whistles in the chimney and the rain pelts the roof? But, of all God's creation, what is so miserable to behold or to think of as a wretched, half-starved family creeping to their nest of flocks or straw, there to lie shivering, till sent forth by the fear of absolutely expiring from want?

HOPS

202. I TREATED of them before; but before I conclude this little Work, it is necessary to speak of them again. I made a mistake as to the tax on the Hops. The positive tax is 2d. a pound, and I (in former editions) stated it at 4d. However, in all such

cases, there falls upon the consumer the expenses attending the paying of the tax. That is to say, the cost of interest of capital in the grower who pays the tax, and who must pay for it, whether his hops be cheap or dear. Then the trouble it gives him, and the rules he is compelled to obey in the drying and bagging, and which cause him great expense. So that the tax on hops of our own English growth, may now be reckoned to cost the consumer about $3\frac{1}{4}$d. a pound.

YEAST

203. YEAST is a great thing in domestic management. I have once before published a receipt for making yeast-cakes. I will do it again here.

204. In Long Island they make yeast-cakes. A parcel of these cakes is made once a year. That is often enough. And, when you bake, you take one of these cakes (or more, according to the bulk of the batch), and with them raise your bread. The very best bread I ever ate in my life was lightened with these cakes.

205. The materials for a good batch of cakes are as follows:—3 ounces of good fresh Hops; $3\frac{1}{2}$ pounds of Rye-Flour; 7 pounds of Indian Corn Meal; and one Gallon of Water. Rub the hops so as to separate them. Put them into the water, which is to be boiling at the time. Let them boil half an hour. Then strain the liquor through a fine sieve into an earthen vessel. While the liquor is hot, put in the Rye-Flour;

stirring the liquor well, and quickly, as the Rye-Flour goes into it. The day after, when it is working, put in the Indian Meal, stirring it well as it goes in. Before the Indian Meal be all in, the mess will be very stiff; and it will, in fact, be dough, very much of the consistence of the dough that bread is made of. Take this dough; knead it well, as you would for pie-crust. Roll it out with a rolling-pin, as you roll out pie-crust, to the thickness of about a third of an inch. When you have it (or part of it at a time) rolled out, cut it up into cakes with a tumbler glass turned upside down, or with something else that will answer the same purpose. Take a clean board (a tin may be better) and put the cakes to dry in the sun. Turn them every day; let them receive no wet; and they will become as hard as ship biscuit. Put them into a bag, or box, and keep them in a place perfectly free from damp. When you bake, take two cakes, of the thickness above-mentioned, and about 3 inches in diameter; put them into hot water, over-night, having cracked them first. Let the vessel containing them stand near the fire-place all night. They will dissolve by the morning, and then you use them in setting your sponge (as it is called) precisely as you would use the yeast of beer.

206. There are two things which may be considered by the reader as obstacles. FIRST, where are we to get the Indian meal? Indian Meal is used merely because it is of a less adhesive nature than that of wheat. White pea-meal, or even barley-meal, would do just as well. But SECOND, to dry the cakes, to make them (and quickly, too, mind) as hard as

ship biscuit (which is much harder than the timber of Scotch firs or Canada firs); and to do this in the sun (for it must not be fire), where are we, in this climate, to get the sun? In 1816 we could not; for, that year, melons rotted in the glazed frames and never ripened. But in every nine summers out of ten, we have, in June, in July, or in August, a fortnight of hot sun, and that is enough. Nature has not given us a peach climate; but we get peaches. The cakes, when put in the sun, may have a glass-sash, or a hand-light, put over them. This would make their berth hotter than that of the hottest open-air situation in America. In short, to a farmer's wife, or any good housewife, all the little difficulties to the attainment of such an object would appear as nothing. The will only is required; and, if there be not that, it is useless to think of the attempt.

SOWING SWEDISH TURNIP SEED

207. It is necessary to be a little more full than I have been before as to the manner of sowing this seed; and I shall make my directions such as to be applied on a small or a large scale. Those that want to transplant on a large scale will of course, as to the other parts of the business, refer to my larger work. It is to get plants for transplanting that I mean to sow the Swedish Turnip Seed. The time for sowing must depend a little upon the nature of the situation and soil. In the north of England, perhaps early in April may be best, but in any of these southern

countries, any time after the middle of April and
before the 10th of May, is quite early enough. The
ground which is to receive the seed should be made
very fine, and manured with wood-ashes, or with
good compost well mixed with the earth. Dung is
not so good, for it breeds the fly more, or at least I
think so. The seed should be sown in drills an inch
deep, made as pointed out under the head of Sowing
in my book on Gardening. When deposited in the
drills evenly but not thickly, the ground should be
raked across the drills so as to fill them up; and then
the whole of the ground should be trodden hard,
with shoes not nailed, and not very thick in the sole.
The ground should be laid out in four feet beds for
the reasons mentioned in the " Gardener." When
the seeds come up, thin the plants to two inches
apart as soon as you think them clear from the fly;
for if left thicker, they injure each other even in this
infant state. Hoe frequently between the rows even
before thinning the plants; and when they are
thinned, hoe well and frequently between them,
for this has a tendency to make them strong, and
the hoeing before thinning helps to keep off the fly.
A rod of ground, the rows being eight inches apart,
and plants two inches apart in the row, will contain
about two thousand two hundred plants. An acre
in rows four feet apart and the plants a foot apart
in the row, will take about ten thousand four hundred
and sixty plants. So that to transplant an acre you
must sow about five rods of ground. The plants
should be kept very clean; and by the last week in
June, or first in July, you put them out. I have put

them out (in England) at all times between the 7th of June and middle of August. The first is certainly earlier than I like; and the very finest I ever grew in England, and the finest I ever saw for a large piece, were transplanted on the 14th of July. But one year with another, the last week in June is the best time.—For size of plants, manner of transplanting, intercultivation, preparing the land, and the rest, see " Year's Residence in America."

No. VIII

On the converting of English Grass, and Grain Plants, cut green, into Straw, for the purpose of making Plat for Hats and Bonnets.

Kensington, May 30, 1823.

208. The foregoing Numbers have treated chiefly of the management of the affairs of a labourer's family, and more particularly of the mode of disposing of the money earned by the labour of the family. The present Number will point out what I hope may become an advantageous kind of labour. All along I have proceeded upon the supposition that the wife and children of the labourer be, as constantly as possible, employed in work of some sort or other. The cutting, the bleaching, the sorting, and the platting of straw, seem to be, of all employments, the best suited to the wives and children of country labourers; and the discovery which I have

made, as to the means of obtaining the necessary materials, will enable them to enter at once upon that employment.

209. Before I proceed to give my directions relative to the performance of this sort of labour, I shall give a sort of history of the discovery to which I have just alluded.

210. The practice of making hats, bonnets, and other things, of straw, is perhaps of very ancient date; but not to waste time in fruitless inquiries, it is very well known that, for many years past, straw coverings for the head have been greatly in use in England, in America, and indeed in almost all the countries that we know much of. In this country the manufacture was, only a few years ago very flourishing; but it has now greatly declined, and has left in poverty and misery those whom it once well fed and clothed.

211. The cause of this change has been, the importation of the straw hats and bonnets from Italy, greatly superior, in durability and beauty, to those made in England. The plat made in England was made of the straw of ripened grain. It was in general split; but the main circumstance was, that it was made of the straw of ripened grain; while the Italian plat was made of the straw of grain or grass cut green. Now, the straw of ripened grain or grass is brittle, or, rather, rotten. It dies while standing, and in point of toughness, the difference between it and straw from plants cut green is much about the same as the difference between a stick that has died on the tree, and one that has been cut from the tree. But be-

sides the difference in point of toughness, strength, and durability, there was the difference in beauty. The colour of the Italian plat was better; the plat was brighter; and the Italian straws, being small whole straws, instead of small straws made by the splitting of large ones, there was a roundness in them that gave light and shade to the plat, which could not be given by our flat bits of straw.

212. It seems odd that nobody should have set to work to find out how the Italians came by this fine straw. The importation of these Italian articles was chiefly from the port of LEGHORN, and therefore the bonnets imported were called Leghorn Bonnets. The straw manufacturers in this country seem to have made no effort to resist this invasion from Leghorn. And, which is very curious, the Leghorn straw has now begun to be imported, and to be platted in this country. So that we had hands to plat as well as the Italians. All that we wanted was the same kind of straw that the Italians had: and it is truly wonderful that these importations from Leghorn should have gone on increasing year after year, and our domestic manufacture dwindling away at a like pace, without there having been any inquiry relative to the way in which the Italians got their straw! Strange, that we should have imported even straw from Italy without enquiring whether similar straw could not be got in England! There really seems to have been an opinion, that England could no more produce this straw than it could produce the sugar-cane.

213. Things were in this state, when in 1821 a

Miss WOODHOUSE, a farmer's daughter in CONNEC-
TICUT, sent a straw bonnet of her own making to the
Society of Arts in London. This bonnet, superior
in fineness and beauty to anything of the kind that
had come from Leghorn, the maker stated to consist of a sort of grass of which she sent along with
the bonnet some of the seeds. The question was,
then, would these precious seeds grow and produce
plants in perfection in England? A large quantity
of the seed had not been sent; and it was therefore,
by a member of the Society, thought desirable to
get, with as little delay as possible, a considerable
quantity of the seed.

214. It was in this stage of the affair that my
attention was called to it. The member just alluded
to applied to me to get the seed from America. I
was of opinion that there could be no sort of grass
in Connecticut that would not, and that did not,
grow and flourish in England. My son JAMES, who
was then at New York, had instructions from me,
in June 1821, to go to Miss WOODHOUSE, and to send
me home an account of the matter. In September
the same year I heard from him, who sent me an
account of the cutting and bleaching, and also a
specimen of the plat and grass of Connecticut. Miss
WOODHOUSE had told the Society of Arts, that the
grass used was the Poa Pratensis. This is the smooth-
stalked meadow-grass. So that it was quite useless
to send for seed. It was clear, that we had grass enough
in England, if we could but make it into straw as
handsome as that of Italy.

215. Upon my publishing an account of what

had taken place with regard to the Amercian Bonnet, an Importer of Italian straw applied to me to know whether I would undertake to import American straw. He was in the habit of importing Italian straw, and of having it platted in this country; but having seen the bonnet of MISS WOODHOUSE, he was anxious to get the American straw. This gentleman showed me some Italian straw which he had imported, and as the seed heads were not on, he could not see what plant it was. The gentleman who showed the straw to me, told me (and doubtless he believed) that the plant was one that would not grow in England. I however, who looked at the straw with the eyes of a farmer, perceived that it consisted of dry oat, wheat, and rye plants, and of Bennet and other common grass plants.

216. This quite settled the point of growth in England. It was now certain that we had the plants in abundance, and the only question that remained to be determined was, Had we SUN to give to those plants the beautiful colour which the American and Italian straw had? If that colour were to be obtained by art, by any chemical applications, we could obtain it as easily as the Americans or the Italians; but if it were the gift of the SUN solely, here might be a difficulty impossible for us to overcome. My experiments have proved that the fear of such difficulty was wholly groundless.

217. It was late in September, 1821, that I obtained this knowledge, as to the kind of plants that produced the foreign straw. I could, at that time of the year, do nothing in the way of removing my

doubts as to the powers of our Sun in the bleaching of grass; but I resolved to do this when the proper season for bleaching should return. Accordingly, when the next month of June came, I went into the country for the purpose. I made my experiments, and in short I proved to demonstration, that we had not only the plants, but the sun also, necessary for the making of straw, yielding in no respect to that of America or Italy. I think that, upon the whole, we have greatly the advantage of those countries, for grass is more abundant in this country than in any other. It flourishes here more than in any other country. It is here in a greater variety of sorts, and for fineness in point of size, there is no part of the world which can equal what might be obtained from some of our downs, merely by keeping the land ungrazed till the month of July.

218. When I had obtained the straw, I got some of it made into plat. One piece of this plat was equal in point of colour, and superior in point of fineness, even to the plat of the bonnet of Miss WOODHOUSE. It seemed, therefore, now to be necessary to do nothing more than to make all this well known to the country. As the SOCIETY OF ARTS had interested itself in the matter, and as I heard that, through its laudable zeal, several sowings of the foreign grass-seed had been made in England, I communicated an account of my experiments to that Society. The first communication was made by me on the 19th of February last, when I sent to the Society specimens of my straw and also of the plat. Some time after this I attended a committee of the Society on the

subject, and gave them a verbal account of the way in which I had gone to work.

219. The committee had before this given some of my straw to certain manufacturers of plat, in order to see what it would produce. These manufacturers, with the exception of one, brought such specimens of plat as to induce, at first sight, any one to believe that it was nonsense to think of bringing the thing to any degree of perfection! But, was it possible to believe this? Was it possible to believe that it could answer to import straw from Italy, to pay a twenty per cent. duty on that straw, and to have it platted here; and that it would not answer to turn into plat straw of just the same sort grown in England? It was impossible to believe this; but possible enough to believe that persons now making profit by Italian straw, or plat, or bonnets, would rather that English straw should not come to shut out the Italian and put an end to the Leghorn trade.

220. In order to show the character of the reports of those manufacturers, I sent some parcels of straw into Hertfordshire, and got back, in the course of five days, fifteen specimens of plat. These I sent to the Society of Arts on the 3rd of April; and I here insert a copy of the letter which accompanied them.

TO THE SECRETARY OF THE SOCIETY OF ARTS

Kensington, April 3, 1823.

SIR,—With this letter I send you sixteen specimens of plat, and also eight parcels of straw, in order to show the sorts that the plat is made out of. The numbers of the plat correspond with those of the straw, but each parcel of straw has two numbers attached to it, except in the case of the first number, which is the wheat straw. Of each kind of straw a parcel of the stoutest and a parcel of the smallest were sent to be platted; so that each parcel of the straw now sent, except that of the wheat, refers to two of the pieces of plat. For instance, 2 and 3 of the plat is of the sort of straw marked 2 and 3; 4 and 12 of the plat is of the sort of straw marked 4 and 12; and so on. These parcels of straw are sent in order that you may know the kind of straw, or rather of grass, from which the several pieces of plat have been made. This is very material; because it is by those parcels of straw that the kinds of grass are to be known.

The piece of plat No. 16 is American; all the rest are from my straw. You will see, that 15 is the finest plat of all. No. 7 is from the stout straw of the same kind as No. 15. By looking at the parcel of straw Nos. 7 and 15, you will see what sort of grass this is. The next, in point of beauty and fineness combined, are the pieces Nos. 13 and 8; and by looking at the parcel of straw Nos. 13 and 8, you will see what sort of grass that is. Next comes 10 and 5, which are very beautiful too; and the sort of grass, you will see, is the common Bennet. The wheat, you see, is too coarse; and the rest of the sorts are either too hard or too brittle. I beg you to look at Nos. 10 and 5. Those appear to me to be the thing to supplant the Leghorn. The colour is good, the straws work well, they afford a great variety of sizes, and they come from the common Bennet grass, which grows all over the kingdom, which is cultivated in all our fields, which is in bloom in the fair month of June, which may be grown as fine or as coarse as we please, and ten acres of which would, I dare say, make ten thousand bonnets. However, 7 and 15, and 8 and 13, are very good; and they are to be got in every part of the kingdom.

As to platters, it is to be too childish to believe that they are not to be got, when I could send off these straws, and get back the plat, in the course of five days. Far better work than this would have been obtained if I could have gone on the errand myself. What, then, will people not do, who regularly undertake the business for their livelihood?

I will, as soon as possible, send you an account of the manner in which I went to work with the grass. The card of plat, which I sent you some time ago, you will be so good as to give me back again some time; because I have now not a bit of the American plat left.   I am, Sir,

Your most humble and

Most obedient servant,

WM. COBBETT.

221. I should observe, that these written communications of mine to the SOCIETY, belong, in fact, to it, and will be published in its PROCEEDINGS, a volume of which comes out every year; but, in this case, there would have been a year lost to those who may act in consequence of these communications being made public. The grass is to be got, in great quantities and of the best sorts, only in June and July; and the Society's volume does not come out till December. The Society has, therefore, given its consent to the making of the communications public through the means of this little work of mine.

222. Having shown what sort of plat could be produced from English grass-straw, I next communicated to the Society an account of the method which I pursued in the cutting and bleaching of the grass. The letter in which I did this I shall here insert a copy of before I proceed further. In the original the paragraphs were numbered from one to seventeen: they are here marked by letters, in order to avoid confusion, the paragraphs of the work itself being marked by numbers.

TO THE SECRETARY OF THE SOCIETY OF ARTS

Kensington, April 14, 1823.

A.—SIR,—Agreeably to your request, I now communicate to you a statement of those particulars which you wished to possess, relative to the specimens of straw and of plat which I have at different times sent to you for the inspection of the Society.

B.—That my statement may not come too abruptly upon those members of the Society who have not had an opportunity of witnessing the progress of this interesting inquiry, I will take a short review of the circumstances which led to the making of my experiments.

C.—In the month of June, 1821, a gentleman, a member of the Society, informed me, by letter, that a Miss WOODHOUSE, a farmer's daughter, of Weathersfield, in Connecticut, had transmitted to the Society a straw bonnet of very fine materials and manufacture; that this bonnet (according to her account) was made from the straw of a sort of grass called Poa Pratensis; that it seemed to be unknown whether the same grass would grow in England; that it was desirable to ascertain whether this grass would grow in England; that, at all events, it was desirable to get from America some of the seed of this grass; and that for this purpose, my informant, knowing that I had a son in America, addressed himself to me, it being his opinion, that, if materials similar to those used by Miss WOODHOUSE could by any means be grown in England, the benefit to the nation must be considerable.

D.—In consequence of this application, I wrote to my son James (then at New York), directing him to do what he was able in order to cause success to the undertaking. On the receipt of my letter, in July, he went from New York to Weathersfield (about a hundred and twenty miles); saw Miss WOODHOUSE;

made the necessary inquiries; obtained a specimen of the grass, and also the plat, which other persons at Weathersfield, as well as Miss WOODHOUSE, were in the habit of making; and having acquired the necessary information as to cutting the grass and bleaching the straw, he transmitted to me an account of the matter; which account, together with his specimens of grass and plat, I received in the month of September.

E.—I was now, when I came to see the specimen of grass, convinced that Miss WOODHOUSE's materials could be grown in England; a conviction which, if it had not been complete at once, would have been made complete immediately afterwards by the sight of a bunch of bonnet straw imported from Leghorn, which straw was shown to me by the importer, and which I found to be that of two or three sorts of our common grass, and of oats, wheat, and rye.

F.—That the grass, or plants, could be grown in England, was, therefore, now certain, and indeed that they were in point of commonness next to the earth itself. But before the grass could, with propriety, be called materials for bonnet making, there was the bleaching to be performed, and it was by no means certain that this could be accomplished by means of an English sun, the difference between which and that of Italy or Connecticut was well known to be very great.

G.—My experiments have, I presume, completely removed this doubt. I think that the straw produced by me to the Society, also some of the pieces of plat, are of a colour which no straw or plat can surpass.

All that remains, therefore, is for me to give an account of the manner in which I cut and bleached the grass which I have submitted to the Society in the state of straw.

H.—First, as to the season of the year, all the straw, except that of one sort of couch-grass, and the long coppice-grass, which two were got in Sussex, were got from grass cut in Hertfordshire on the 21st of June. A grass head-land, in a wheat-field, had been mowed during the fore-part of the day, and in the afternoon I went and took a handful here and a handful there out of the swaths. When I had collected as much as I could well carry, I took it to my friend's house, and proceeded to prepare it for bleaching, according to the information sent me from America by my son; that is to say, I put my grass into a shallow tub, put boiling water upon it until it was covered with water, let it remain in that state for ten minutes, then took it out, and laid it very thinly on a closely-mowed lawn in a garden. But I should observe, that, before I put the grass into the tub, I tied it up in small bundles, or sheaves, each bundle being about six inches through at the butt-end. This was necessary, in order to be able to take the grass, at the end of ten minutes, out of the water, without throwing it into a confused mixture as to tops and tails. Being tied up in little bundles, I could easily, with a prong, take it out of the hot water. The bundles were put into a large wicker basket, carried to the lawn in the garden, and there taken out, one by one, and laid in swaths as before mentioned.

I.—It was laid very thinly; almost might I say,

that no stalk of grass covered another. The swaths were turned once a day. The bleaching was completed at the end of seven days from time of scalding and laying out. June is a fine month. The grass was, as it happened, cut on the longest day in the year; and the weather was remarkably fine and clear. But the grass which I afterwards cut in Sussex, was cut in the first week in August; and as to the weather my journal speaks thus:—

August, 1822.

2nd.—Thunder and rain.—Began cutting grass.

3rd.—Beautiful day.

4th.—Fine day.

5th.—Cloudy day.—Began scalding grass, and laying it out.

6th.—Cloudy greater part of the day.

7th.—Same weather.

8th.—Cloudy and rather misty.—Finished cutting grass.

9th.—Dry, but cloudy.

10th.—Very close and hot.—Packed up part of the grass.

11th.—Same weather.

12th.
13th. } Same weather.
14th.

15th.—Hot and clear.—Finished packing the grass.

K.—The grass cut in Sussex was as well bleached as that cut in Hertfordshire; so that it is evident that we never can have a summer that will not afford sun sufficient for this business.

L.—The part of the straw used for platting is that part of the stalk which is above the upper joint; that part which is between the upper joint and the seed-branches. This part is taken out, and the rest of the straw thrown away. But the whole plant must be cut and bleached; because, if you were to take off, when green, the part above described, that part would wither up next to nothing. This part must die in company with the whole plants, and be separated from the other parts after the bleaching has been performed.

M.—The time of cutting must vary with the seasons, the situation, and the sort of grass. The grass which I got in Hertfordshire, than which nothing can, I think, be more beautiful, was, when cut, generally in bloom; just in bloom. The wheat was in full bloom; so that a good time for getting grass may be considered to be that when the wheat is in bloom. When I cut the grass in Sussex, the wheat was ripe, for reaping had begun; but that grass is of a very backward sort, and, besides, grew in the shade amongst coppice-wood and under trees, which stood pretty thick.

N.—As to the sorts of grass, I have to observe generally, that in proportion as the colour of the grass is deep; that is to say, getting further from the yellow and nearer to the blue, it is of a deep and dead yellow when it becomes straw. Those kinds of grass are best which are, in point of colour, nearest to that of wheat, which is a fresh pale green. Another thing is, the quality of the straw as to pliancy and toughness. Experience must be our guide here. I had not time to make a large collection of sorts; but those which I have sent to you contain three sorts which are proved to be good. In my letter of the 3rd instant I sent you sixteen pieces of plat and eight bunches of straw, having the seed heads on, in order to show the sorts of grass. The sixteenth piece of plat was American. The first piece was from wheat cut and bleached by me; the rest from grass cut and bleached by me. I will here, for fear of mistake, give a list of the names of the several sorts of grass, the straw of which was sent with my letter of the 3rd instant, referring to the numbers, as placed on the plat and on the bunches of straw.

PIECES OF PLAT.	BUNCHES OF STRAW.	SORTS OF GRASS.
No. 1	No. 1	Wheat.
2. 3.	2 and 3	Melica Cærulea, or Purple Melica Grass.
4. 12.	4 and 12	Agrostis Stolonifera, or Fiorin Grass; that is to say, one sort of Couch-Grass.
5. 10.	5 and 10	Lolium Perenne, or Ray-Grass.
6. 11.	6 and 11	Avena Flavescens, or Yellow Oat-Grass.
7. 15.	7 and 15	Cynosurus Cristatus, or Crested Dog's-tail Grass.
8. 13.	8 and 13	Anthoxanthum Oderatum, or Sweet-scented Vernal Grass.
9. 14.	9 and 14	Agrostis Canina, or Brown Bent Grass.

O.—These names are those given at the Botanical Garden at Kew. But the same English names are not in the country given to these sorts of grass. The Fiorin grass, the Yellow Oat-grass, and the Brown-Bent, are all called Couch-grass; except that the latter is, in Sussex, called Red Robin. It is the native grass of the plains of Long Island, and they call it Red Top. The Ray-grass is the common field-grass, which is, all over the kingdom, sown with clover. The farmers, in a great part of the kingdom, call it Ben't, or Bennet grass, and sometimes it is called Darnel-grass. The Crested Dog's-tail goes in Sussex by the name of Hendonbent, for what reason I know not. The Sweet-scented Vernal-grass I have never, amongst the farmers, heard any name for. Miss Wood-house's grass appears, from the plants that I saw in the Adelphi, to be one of the sorts of Couch-grass. Indeed, I am sure that it is a Couch-grass, if the plants I there saw came from her seed. My son, who went into Connecticut, who saw the grass grow-ing, and who sent me home a specimen of it, is now in England; he was with me when I cut the grass in Sussex, and he says that Miss Woodhouse's was a Couch-grass. However, it is impossible to look at the specimens of straw and of plat which I have sent you, without being convinced that there is no want of the raw material in England. I was, after my first hearing of the subject, very soon convinced that the grass grew in England; but I had great doubts as to the capacity of our sun. Those doubts my own ex-periments have completely removed, but then I was not aware of the great effect of the scalding, of

which, by the way, Miss WOODHOUSE had said nothing, and the knowledge of which we owe entirely to
my son James's journey into Connecticut.

P.—Having thus given you an account of the time
and manner of cutting the grass, of the mode of
cutting and bleaching; having given you the best
account I am able, as to the sorts of grass to be
employed in this business; and having, in my former
communications, given you specimens of the plat
wrought from the several sorts of straw, I might
here close my letter; but, as it may be useful to speak
of the expense of cutting and bleaching, I shall
trouble you with a few words relating to it. If there
were a field of Ray-grass, or of Crested Dog's-tail,
or any other good sort, and nothing else growing
with it, the expense of cutting would be very little
indeed, seeing that the scythe or reap-hook would
do the business at a great rate. Doubtless there will
be such fields; but even if the grass have to be cut
by the handful, my opinion is, that the expense of
cutting and bleaching would not exceed fourpence
for straw enough to make a large bonnet. I should
be willing to contract to supply straw at this rate for
half a million of bonnets. The scalding must con-
stitute a considerable part of the expense, because
there must be fresh water for every parcel of grass
that you put in the tub. When water has scalded one
parcel of cold grass, it will not scald another parcel.
Besides, the scalding draws out the sweet matter of
the grass, and makes the water the colour of that
horrible stuff called London porter. It would be
very good, by-the-by, to give to pigs. Many people

give hay-tea to pigs and calves, and this is grass-tea. To scald a large quantity, therefore, would require means not usually at hand, and the scalding is an essential part of the business. Perhaps, in a large and convenient farm-house, with a good brewing copper, good fuel and water handy, four or five women might scald a wagon-load in a day; and a wagon-load would, I think, carry straw enough (in the rough) to furnish the means of making a thousand bonnets. However, the scalding might take place in the field itself, by means of a portable boiler, especially if water were at hand; and perhaps it would be better to carry the water to the field than to carry the grass to the farm-house, for there must be ground to lay it out upon the moment it has been scalded, and no ground can be so proper as the newly-mowed ground where the grass has stood. The space, too, must be large for any considerable quantity of grass. As to all these things, however, the best and cheapest methods will soon be discovered when people set about the work with a view to profit.

Q.—The Society will want nothing from me, nor from anybody else, to convince it of the importance of this matter; but I cannot, in concluding these communications to you, Sir, refrain from making an observation or two on the consequences likely to arise out of these inquiries. The manufacture is alone of considerable magnitude. Not less than about five millions of persons in this kingdom have a dress which consists partly of manufactured straw; and a large part, and all the most expensive part, of the articles thus used now come from abroad. In cases where you can get from abroad any article at less expense than you can get it at home, the wisdom of fabricating that article at home may be doubted. But in this case you get the raw material by labour performed at home, and the cost of that labour is not nearly so great as would be the cost of the mere carriage of the straw from a foreign country to this. If our own people had all plenty of employment, and that, too, more profitable to them and to the country than the turning of a part of our own grass into articles of dress, then it would be advisable still to import Leghorn bonnets; but the facts being the reverse, it is clear that whatever money, or money's worth of things, be sent out of the country, in exchange for Leghorn bonnets, is, while we have the raw material here for next to nothing, just so much thrown away. The Italians, it may be said, take some of our manufactures in exchange; and let us suppose, for the purpose of illustration, that they take cloth from Yorkshire; stop the exchange between Leghorn and Yorkshire, and does Yorkshire lose part of its custom? No: for though those who make the bonnets out of English grass prevent the Leghorners from buying Yorkshire cloth, they, with the money which they now get, instead of its being got by the Leghorners, buy the Yorkshire cloth themselves; and they wear this cloth, too, instead of its being worn by the people of Italy: ay, Sir, and many now in rags will be well clad, if the laudable object of the Society be effected. Besides this, however, why should we not export the articles of this manufacture? To America we certainly should; and I should not be at all surprised if we were to export them to Leghorn itself.

R.—Notwithstanding all this, however, if the manufacture were of a description to require, in order to give it success, the collecting of the manufacturers together in great numbers, I should, however great the wealth that it might promise, never have done anything to promote its establishment. The contrary is happily the case: here all is not only performed by hand, but by hand singly, without any combination of hands. Here there is no power of machinery or of chemistry wanted. All is performed out in the open fields, or sitting in the cottage. There want no coal mines and no rivers to assist, no water-powers nor powers of fire. No part of the kingdom is unfit for the business. Everywhere there are grass, water, sun, and women and children's fingers; and these are all that are wanted. But the great thing of all is this: that, to obtain the materials for the making of this article of dress, at once so gay, so useful, and in some cases so expensive, there requires not a penny of capital. Many of the labourers now make their own straw hats to wear in summer. Poor rotten things, made out of straw of ripened grain. With what satisfaction will they learn that straw, twenty times as durable, to say nothing of the beauty, is to be got from every hedge? In short, when the people are well and clearly informed of the facts, which I have through you, Sir, had the honour to lay before the Society, it is next to impossible that the manufacture should not become general throughout the country. In every labourer's house a pot of water can be boiled. What labourer's wife cannot, in the summer months, find time to cut and bleach grass enough to give her and her children work for a part of the winter? There is no necessity for all to be platters. Some may cut and bleach only. Others may prepare the straw, as mentioned in paragraph L. of the letter. And doubtless, as the farmers in Hertfordshire now sell their straw to the platters, grass collectors and bleachers and preparers would do the same. So that there is scarcely any country labourer's family that might not derive some advantage from this discovery; and while I am convinced that this consideration has been by no means overlooked by the Society, it has been, I assure you, the great consideration of all with,

Sir, your most obedient and

Most humble servant,

WM. COBBETT.

223. In the last edition, this closing part of the work, relative to the straw plat, was not presented to the public as a thing which admitted of no alteration; but on the contrary, it was presented to the public with the following concluding remark:—
" In conclusion, I have to observe, that I by no
" means send forth this essay as containing opinions
" and instructions that are to undergo no alteration.
" I am, indeed, endeavouring to teach others, but I
" am myself only a learner. Experience will, doubt-
" less, make me much more perfect in a knowledge
" of the several parts of the subject; and the fruit

" of this experience I shall be careful to communi-
" cate to the public." I now proceed to make good
this promise. Experience has proved that very beautiful and very fine plat can be made of the straw of
divers kinds of grass. But the most ample experience
has also proved to us that it is to the straw of wheat
that we are to look for a manufacture to supplant
the Leghorn. This was mentioned as a strong sus-
picion in my former edition of this work. And I
urged my readers to sow wheat for the purpose.
The fact is now proved beyond all contradiction
that the straw of wheat or rye, but particularly of
wheat, is the straw for this purpose. Finer plat may
be made from the straw grass than can possibly be
made from the straw of wheat or rye: but the grass
plat is, all of it, more or less brittle; and none of it
has the beautiful and uniform colour of the straw of
wheat. Since the last edition of this work, I have
received packets of the straw from Tuscany, all of
wheat; and indeed I am convinced that no other
straw is anything like so well calculated for the purpose. Wheat straw bleaches better than any other. It
has that fine pale, golden colour which no other
straw has; it is much more simple, more pliant than
any other straw; and, in short, this is the material.
I did not urge in vain. A good quantity of wheat was
sowed for this purpose. A great deal of it has been
well harvested; and I have the pleasure to know that
several hundreds of persons are now employed in
the platting of straw. One more year, one more
crop of wheat, and another Leghorn bonnet will
never be imported into England. Some great errors

have been committed in the sowing of the wheat and in the cutting of it. I shall now, therefore, availing myself of the experience which I have gained, offer to the public some observations on the sort of wheat to be sowed for this purpose; on the season for sowing; on the land to be used for the purpose; on the quantity of seed and the manner of sowing; on the season for cutting; on the manner of cutting, bleaching, and housing; on the platting; on the knitting; and on the pressing.

224. The SORT OF WHEAT. The Leghorn plat is all made of the straw of the spring wheat. This spring wheat is so called by us, because it is sowed in the spring, at the same time that barley is sowed. The botanical name of it is TRITICUM ÆSTIVUM. It is a small-grained bearded wheat. It has very fine straw; but experience has convinced me, that the little brown-grained winter wheat is just as good for the purpose. In short, any wheat will do. I have now in my possession specimens of plat made of both winter and spring wheat, and I see no difference at all. I am decidedly of opinion that the winter wheat is as good as the spring wheat for the purpose. I have plat and I have straw both now before me, and the above is the result of my experience.

225. THE LAND PROPER FOR THE GROWING OF WHEAT. The object is to have the straw as small as we can get it. The land must not, therefore, be too rich; yet it ought not to be very poor. If it be, you get the straw of no length. I saw an acre this year, as beautiful as possible, sowed upon a light

loam, which bore last year a fine crop of potatoes. The land ought to be perfectly clean, at any rate; so that, when the crop is taken off, the wheat straw may not be mixed with weeds and grass.

226. SEASON FOR SOWING. This will be more conveniently stated in paragraph 228.

227. QUANTITY OF SEED AND MANNER OF SOWING. When first this subject was started in 1821, I said, in the Register, that I would engage to grow as fine straw in England as the Italians could grow. I recommended, then, as a first guess, fifteen bushels of wheat to the acre. Since that, reflection told me that that was not quite enough. I therefore recommend twenty bushels to the acre. Upon the beautiful acre which I have mentioned above, eighteen bushels, I am told, were sowed; fine and beautiful as it was, I think it would have been better if it had had twenty bushels; twenty bushels, therefore, is what I recommend. You must sow broad-cast, of course, and you must take great pains to cover the seed well. It must be a good even-handed seedsman, and there must be very nice covering.

228. SEASON FOR CUTTING. Now, mind, it is fit to cut in just about one week after the bloom is dropped. If you examine the ear at that time, you will find the grain just beginning to be formed, and that is precisely the time to cut the wheat. The straw has then got its full substance in it. But I must now point out a very material thing. It is by no means desirable to have all your wheat fit to cut at the same time. It is a great misfortune, indeed, so to have it. If fit to cut altogether, it ought to be cut all at the

same time; for supposing you to have an acre, it will require a fortnight or three weeks to cut it and bleach it, unless you have a very great number of hands, and very great vessels to prepare water in. Therefore, if I were to have an acre of wheat for this purpose, and were to sow all spring wheat, I would sow a twelfth part of the acre every week from the first week in March to the last week in May. If I relied partly upon winter wheat, I would sow some every month, from the latter end of September to March. If I employed the two sorts of wheat, or, indeed, if I employed only the spring wheat, the TRITICUM ÆSTIVUM, I should have some wheat fit to cut in June, and some not fit to cut till September. I should be sure to have a fair chance as to the weather. And, in short, it would be next to impossible for me to fail of securing a considerable part of my crop. I beg the reader's particular attention to the contents of this paragraph.

229. MANNER OF CUTTING THE WHEAT. It is cut by a little reap-hook, close to the ground as possible. It is then tied in little sheaves, with two pieces of string, one near the butt, and the other about half-way up. This little bundle or sheaf ought to be six inches through at the butt and no more. It ought not to be tied too tightly, lest the scalding should not be perfect.

230. MANNER OF BLEACHING. The little sheaves mentioned in the last paragraph are carried to a brewing mash, vat, or other tub. You must not put them into the tub in too large a quantity, lest the water get chilled before it get to the bottom.

Pour on scalding water till you cover the whole of the little sheaves, and let the water be a foot above the top sheaves. When the sheaves have remained thus a full quarter of an hour, take them out with a prong, lay them in a clothes-basket, or upon a hurdle, and carry them to the ground where the bleaching is to be finished. This should be, if possible, a piece of grass land, where the grass is very short. Take the sheaves, and lay some of them along in a row; untie them, and lay the straw along in that row as thin as it can possibly be laid. If it were possible, no one straw ought to have another lying upon it, or across it. If the sun be clear, it will require to lie twenty-four hours thus, then be turned, and lie twenty-four hours on the other side. If the sun be not very clear, it must lie longer. But the numerous sowings which I have mentioned will afford you so many chances, so many opportunities of having fine weather, that the risk about weather would necessarily be very small. If wet weather should come, and if your straw remain out in it any length of time, it will be spoiled; but, according to the mode of sowing above pointed out, you really could stand very little chance of losing straw by bad weather. If you had some straw out bleaching, and the weather were to appear suddenly to be about to change, the quantity that you would have out would not be large enough to prevent you from putting it under cover, and keeping it there till the weather changed.

231. HOUSING THE STRAW. When your straw is nicely bleached, gather it up, and with the same string that you used to tie it when green, tie it

up again into little sheaves. Put it by in some room where there is no damp, and where mice and rats are not suffered to inhabit. Here it is always ready for use, and it will keep, I dare say, four or five years, very well.

232. THE PLATTING. This is now so well understood that nothing need be said about the manner of doing the work. But much might be said about the measures to be pursued by land-owners, by parish officers, by farmers, and more especially by gentlemen and ladies of sense, public spirit, and benevolence of disposition. The thing will be done; the manufacture will spread itself over all this kingdom; but the exertions of those whom I have here pointed out might hasten the period of its being brought to perfection. And I beg such gentlemen and ladies to reflect on the vast importance of such manufacture, which it is impossible to cause to produce anything but good. One of the great misfortunes of England at this day is, that the land has had taken away from it those employments for its women and children which were so necessary to the well-being of the agricultural labourer. The spinning, the carding, the reeling, the knitting; these have been all taken away from the haughty lords of bands of abject slaves, and given to the Lords of the Loom. But let the landholder mark how the change has operated to produce his ruin. He must have the labouring MAN and the labouring BOY; but, alas! he cannot have these, without having the man's wife and the boy's mother, and little sisters and brothers. Even Nature herself says, that he shall have

the wife and little children, or that he shall not have the man and the boy. But the Lords of the Loom, the crabbed-voiced, hard-favoured, hard-hearted, puffed-up, insolent, savage and bloody wretches of the North have, assisted by a blind and greedy Government, taken all the employment away from the agricultural women and children. This manufacture of straw will form one little article of employment for these persons. It sets at defiance all the hatching and scheming of all the tyrannical wretches who cause the poor little creatures to die in their factories, heated to eighty-four degrees. There will need no inventions of WATT; none of your horse powers, nor water powers; no murdering of one set of wretches in the coal mines, to bring up the means of murdering another sort of wretches in the factories, by the heat produced from these coals; none of these are wanted to carry on this manufacture. It wants no combination laws; none of the inventions of the hard-hearted wretches of the North.

233. THE KNITTING. Upon this subject, I have only to congratulate my readers that there are great numbers of English women who can now knit plat together, better than those famous Jewesses of whom we are told.

234. THE PRESSING. Bonnets and hats are pressed after they are made. I am told that a proper press costs pretty nearly a hundred pounds; but, then, that it will do a prodigious deal of business. I would recommend to our friends in the country to teach as many children as they can to make the plat. The plat will be knitted in London, and in other

considerable towns, by persons to whom it will be sold. It appears to me, at least, that this will be the course that the thing will take. However, we must leave this to time: and here I conclude my observations upon a subject which is deeply interesting to myself, and which the public in general deem to be of great importance.

235. POSTSCRIPT on Brewing.—I think it right to say here, that, ever since I published the instructions for brewing by copper and by wooden utensils, the beer at my own house has always been brewed precisely agreeably to the instructions contained in this book; and I have to add, that I never have had such good beer in my house in all my lifetime, as since I have followed that mode of brewing. My table-beer, as well as my ale, is always as clear as wine. I have had hundreds and hundreds of quarters of malt brewed into beer in my house. My people could always make it strong enough and sweet enough; but never, except by accident, could they make it CLEAR. Now I never have any that is not clear. And yet my utensils are all very small; and my brewers are sometimes one labouring man, and sometimes another. A man wants showing how to brew the first time. I should suppose that we use, in my house, about seven hundred gallons of beer every year, taking both sorts together; and I can positively assert, that there has not been one drop of bad beer, and indeed none which has not been most excellent, in my house, during the last two years, I think it is, since I began using the utensils, and in the manner named in this book.

No. IX

ICE-HOUSES

236. First begging the reader to read again paragraph 149, I proceed here, in compliance with numerous requests to that effect, to describe, as clearly as I can, the manner of constructing the sort of Ice-houses therein mentioned. In England, these receptacles of frozen water are, generally, under ground, and always, if possible, under the shade of trees, the opinion being, that the main thing, if not the only thing, is to keep away the heat. The heat is to be kept away certainly; but moisture is the great enemy of Ice; and how is this to be kept away, either under ground, or under the shade of trees? Abundant experience has proved, that no thickness of wall, that no cement of any kind, will effectually resist moisture. Drops will, at all times, be seen hanging on the under side of an arch of any thickness, and made of any materials, if it have earth over it, and even when it has the floor of a house over it; and wherever the moisture enters, the ice will quickly melt.

237. Ice-houses should therefore be, in all their parts, as dry as possible: and they should be so constructed, and the ice so deposited in them, as to ensure the running away of the meltings as quickly as possible, whenever such meltings come. Anything in the way of drains or gutters is too slow in its effect; and therefore there must be something that will not suffer the water, proceeding from any melting, to remain an instant.

238. In the first place, then, the ice-house should stand in a place quite open to the sun and air; for whoever has travelled even but a few miles (having eyes in his head) need not be told how long that part of a road from which the sun and wind are excluded by trees, or hedges, or by anything else, will remain wet, or at least damp, after the rest of the road is even in a state to send up dust.

239. The next thing is to protect the ice against wet, or damp, from beneath. It should, therefore, stand on some spot from which water would run in every direction; and if the natural ground present no such spot, it is no very great job to make it.

240. Then come the materials of which the house is to consist. These, for the reasons before-mentioned, must not be bricks, stones, mortar, or earth; for these are all affected by the atmosphere; they will become damp at certain times, and dampness is the great destroyer of ice. The materials left are wood and straw. Wood will not do; for though not liable to become damp, it imbibes heat fast enough; and, besides, it cannot be so put together as to shut out air sufficiently. Straw is wholly free from the quality of becoming damp, except from water actually put upon it; and it can, at the same time, be placed on a roof, and on sides, to such a degree of thickness as to exclude the air in a manner the most perfect. The ice-house ought, therefore, to be made of posts, plates, rafters, laths, and straw. The best form is the circular; and the house, when made, appears as I have endeavoured to describe it in FIG. 3 of the plate.

241. FIG. 1, *a*, is the centre of a circle, the di-

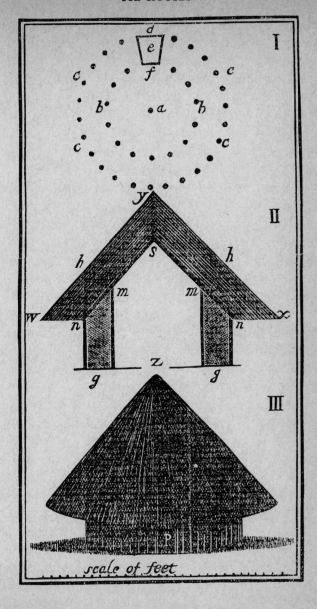

scale of feet

ameter of which is ten feet, and at this centre you put up a post to stand fifteen feet above the level of the ground, which post ought to be about nine inches through at the bottom, and not a great deal smaller at the top. Great care must be taken that this post be perfectly perpendicular; for, if it be not, the whole building will be awry.

242. *b b b* are fifteen posts, nine feet high, and six inches through at the bottom, without much tapering towards the top. These posts stand about two feet apart, reckoning from centre of post to centre of post, which leaves between each two a space of eighteen inches. *c c c c* are fifty-four posts, five feet high, and five inches through at the bottom, without much tapering towards the top. These posts stand about two feet apart, from centre of post to centre of post, which leaves between each two a space of nineteen inches. The space between these two rows of posts is four feet in width, and, as will be presently seen, is to contain a wall of straw.

243. *e* is a passage through this wall; *d* is the out-side door of the passage; *f* is the inside door; and the inner circle, of which *a* is the centre, is the place in which the ice is to be deposited.

244. Well, then, we have now got the posts up; and, before we talk of the roof of the house, or of the bed for the ice, it will be best to speak about the making of the wall. It is to be made of straw, wheat-straw, or rye-straw, with no rubbish in it, and made very smooth by the hand as it is put in. You lay it in very closely and very smoothly, so that if the wall were cut across, as at *g g*, in FIG. 2 (which FIG. 2

represents the whole building cut down through the middle, omitting the centre post), the ends of the straw would present a compact face as they do after a cut of a chaff-cutter. But there requires something to keep the straw from bulging out between the posts. Little stakes as big as your wrist will answer this purpose. Drive them into the ground, and fasten, at top, to the plates, of which I am now to speak. The plates are pieces of wood which go all round both the circles, and are nailed on upon the tops of the posts. Their main business is to receive and sustain the lower ends of the rafters, as at *m m* and *n n*, in FIG. 2. But to the plates also the stakes just mentioned must be fastened at top. Thus, then, there will be this space of four feet wide, having on each side of it a row of posts and stakes, not more than about six inches from each other, to hold up, and to keep in its place, this wall of straw.

245. Next come the rafters, as from *s* to *n*, FIG. 2. Carpenters best know what is the number and what the size of the rafters; but from *s* to *m* there need be only about half as many as from *m* to *n*. However, carpenters know all about this. It is their every-day work. The roof is forty-five degrees pitch, as the carpenters call it. If it were even sharper, it would be none the worse. There will be about thirty ends of rafters to lodge on the plate, as at *m*; and these cannot all be fastened to the top of the centre post rising up from *a*; but carpenters know how to manage this matter, so as to make all strong and safe. The plate which goes along on the tops of the row of posts, *b b b*, must, of course, be put on in a some-

what sloping form; otherwise there would be a sort
of hip formed by the rafters. However, the thatch
is to be so deep, that this may not be of much consequence. Before the thatching begins, there are laths
to put up on the rafters. Thatchers know all about
this, and all that you have to do is, to take care that
the thatcher tie the straw on well. The best way, in
a case of such deep thatch, is to have a strong man
to tie for the thatcher.

246. The roof is now raftered, and it is to receive
a thatch of clean, sound, and well-prepared wheat
or rye straw, four feet thick, as at *h h*, in FIG. 2.

247. The house having now got walls and roof,
the next thing is to make the bed to receive the ice.
This bed is the area of the circle of which *a* is the
centre. You begin by laying on the ground round
logs, eight inches through, or thereabouts, and
placing them across the area, leaving spaces between
them of about a foot, Then, crossways on them,
poles about four inches through, placed at six inches
apart. Then, crossways on them, other poles about
two inches through, placed at three inches apart.
Then, crossways on them, rods as thick as your
finger, placed at an inch apart. Then upon these,
small, clean, dry, last winter-cut twigs, to the thickness of about two inches; or, instead of these twigs,
good, clean, strong, heath free from grass and moss,
and from rubbish of all sorts.

248. This is the bed for the ice to lie on; and, as
you see, the top of the bed will be seventeen inches
from the ground. The pressure of the ice may, perhaps, bring it to fourteen, or to thirteen. Upon this

bed the ice is put, broken and pummelled, and beaten down together in the usual manner.

249. Having got the bed filled with ice, we have next to shut it safely up. As we have seen, there is a passage (*e*). Two feet wide is enough for this passage, and being as long as the wall is thick, it is, of course, four feet long. The use of the passage is this: that you may have two doors, so that you may, in hot or damp weather, shut the outer door, while you have the inner door open. The inner door may be of hurdle-work, and straw, and covered, on one of the sides, with sheep-skins, with the wool on, so as to keep out the external air. The outer door, which must lock, must be of wood made to shut very closely, and, besides, covered with skins, like the other. At times of great danger from heat, or from wet, the whole of the passage may be filled with straw. The door (*p.*, FIG. 3) should face the North, or between North and East.

250. As to the size of the ice-house, that must, of course, depend upon the quantity of ice that you may choose to have. A house on the above scale, is from *w* to *x* (FIG. 2) twenty-nine feet; from *y* to *z* (FIG. 2) nineteen feet. The area of the circle, of which *a* is the centre, is ten feet in diameter, and as this area contains seventy-five superficial feet, you will, if you put ice on the bed to the height of only five feet (and you may put it on to the height of seven feet from the top of the bed), you will have three hundred and seventy-five cubic feet of ice, and observe, a cubic foot of ice will, when broken up, fill much more than a Winchester bushel: what it

may do as to an " IMPERIAL BUSHEL," engendered
like Greek Loan Commissioners, by the unnatural
heat of " PROSPERITY," God only knows! However,
I do suppose, that, without making any allowance
for the " cold fit," as Dr. Baring calls it, into which
the " late panic " has brought us: I do suppose,
that even the scorching, the burning dog-star of
" IMPERIAL PROSPERITY "; nay, that even DIVES
himself, would hardly call for more than two bushels
of ice in a day; for more than two bushels a day it
would be, unless it were used in cold as well as in
hot weather.

251. As to the expense of such a house, it could,
in the country, not be much. None of the posts,
except the main or centre-post, need be very straight.
The other posts might be easily culled from tree-
lops, destined for fire-wood. The straw would make
all straight. The plates must of necessity be short
pieces of wood; and, as to the stakes, the laths, and
the logs, poles, rods, twigs, and heath, they would
not all cost twenty shillings. The straw is the principal article; and, in most places, even that would
not cost more than two or three pounds. If it last
many years, the price could not be an object; and
if but a little while, it would still be nearly as good
for litter as it was before it was applied to this purpose. How often the bottom of the straw walls might
want renewing I cannot say, but I know that the
roof would, with few and small repairs, last well for
ten years.

252. I have said that the interior row of posts is
to be nine feet high, and the exterior row five feet

high. I, in each case, mean, with the plate inclusive. I have only to add, that by way of superabundant precaution against bottom wet, it will be well to make a sort of gutter, to receive the drip from the roof, and to carry it away as soon as it falls.

253. Now, after expressing a hope that I shall have made myself clearly understood by every reader, it is necessary that I remind him, that I do not pretend to pledge myself for the complete success, nor for any success at all, of this mode of making ice-houses. But, at the same time, I express my firm belief, that complete success would attend it; because it not only corresponds with what I have seen of such matters, but I had the details from a gentleman who had ample experience to guide him, and who was a man on whose word and judgment I placed a perfect reliance. He advised me to erect an ice-house; but not caring enough about fresh meat and fish in summer, or at least, not setting them enough above " prime pork " to induce me to take any trouble to secure the former, I never built an ice-house. Thus, then, I only communicate that in which I believe; there is, however, in all cases, this comfort, that if the thing fail as an ice-house, it will serve all generations to come as a model for a pig-bed.

ADDITION

Kensington, Nov. 14, 1831.

MANGEL WURZEL

254. THIS last summer, I have proved, that, as keep for cows, MANGEL WURZEL is preferable to SWEDISH TURNIPS, whether as to quantity or quality. But there needs no other alteration in the Book, than merely to read mangel wurzel wherever you find Swedish turnip; the time of sowing, the mode, and time of transplanting, the distances, and the cultivation, all being the same; and the only difference being in the application of the leaves, and in the time of harvesting the roots.

255. The leaves of the MANGEL WURZEL are of great value, especially in dry summers. You begin, about the third week in August, to take off, by a downward pull, the leaves of the plants; and they are excellent food for pigs and cows; only observe this, that, if given to cows, there must be, for each cow, six pounds of hay a day, which is not necessary in the case of the Swedish turnips. These leaves last till the crop is taken up, which ought to be in the first week of November. The taking off of the leaves does good to the plants: new leaves succeed higher up; and the plant becomes longer than it otherwise

would be, and, of course, heavier. But, in taking off the leaves, you must not approach too near to the top.

256. When you take the plants up in November, you must cut off the crowns and the remaining leaves; and they, again, are for cows and pigs. Then you put the roots into some place to keep them from the frost; and, if you have no place under cover, put them in pies, in the same manner as directed for the Swedish turnips. The roots will average in weight 10 lbs. each. They may be given to cows whole, or to pigs either, and they are better than the Swedish turnip for both animals; and they do not give any bad or strong taste to the milk and butter. But, besides this use of the mangel wurzel, there is another, with regard to pigs at least, of very great importance. The juice of this plant has so much of sweetness in it, that, in France, they make sugar of it; and have used the sugar, and found it equal in goodness to West India sugar. Many persons in England make beer of this juice, and I have drunk of this beer, and found it very good. In short, the juice is most excellent for the mixing of moist food for pigs. I am now (20th Nov., 1831) boiling it for this purpose. My copper holds seven strike-bushels; I put in three bushels of mangel wurzel cut into pieces of two inches thick, and then fill the copper with water. I draw off as much of the liquor as I want to wet pollard, or meal, for little pigs, or fatting-pigs, and the rest, roots and all, I feed the yard-hogs with; and this I shall follow on till about the middle of May.

257. If you give boiled, or steamed, potatoes to pigs, there wants some liquor to mix with the potatoes; for the water in which potatoes have been boiled is hurtful to any animal that drinks it. But mix the potatoes with juice of mangel wurzel, and they make very good food for hogs of all ages. The mangel wurzel produces a larger crop than the Swedish turnip.

COBBETT'S CORN

258. IF you prefer bread and pudding to milk, butter, and meat, this corn will produce, on your forty rods, forty bushels, each weighing 60 lbs. at the least; and more flour, in proportion, than the best white wheat. To make bread with it, you must use two-thirds wheaten, or rye, flour; but in puddings this is not necessary. The puddings at my house are all made with this flour, except meat and fruit pudding; for the corn flour is not adhesive or clinging enough to make paste or crust. This corn is the very best for hog-fatting in the whole world. I, last April, sent parcels of the seed into several counties, to be given away to working men; and I sent them instructions for the cultivation, which I shall repeat here.

259. I will first describe this corn to you. It is that which is sometimes called Indian corn; and sometimes people call it Indian wheat. It is that sort of corn which the disciples ate as they were going up to Jerusalem on the Sabbath-day. They gathered

it in the fields as they went along and ate it green, they being " an hungered," for which, you know, they were reproved by the pharisees. I have written a treatise on this corn in a book which I sell for four shillings, giving a minute account of the qualities, the culture, the harvesting, and the various uses of this corn; but I shall here confine myself to what is necessary for a labourer to know about it, so that he may be induced to raise and may be enabled to raise enough of it in his garden to fat a pig of ten score.

260. There are a great many sorts of this corn. They all come from countries which are hotter than England. This sort, which my eldest son brought into England, is a dwarf kind, and is the only kind that I have known to ripen in this country; and I know that it will ripen in this country in any summer; for I had a large field of it in 1828 and 1829; and last year (my lease at my farm being out at Michaelmas, and this corn not ripening till late in October) I had about two acres in my garden at Kensington. Within the memory of man there have not been three summers so cold as the last, one after another; and no one so cold as the last. Yet my corn ripened perfectly well, and this you will be satisfied of if you be amongst the men to whom this corn is given from me. You will see that it is in the shape of the cone of a spruce fir; you will see that the grains are fixed round a stalk which is called the cob. These stalks or ears come out of the side of the plant, which has leaves like a flag, which plant grows to about three feet high, and has two or three, and sometimes more, of these ears or bunches of grain. Out of the

top of the plant comes the tassel, which resembles the plumes of feathers upon a hearse; and this is the flower of the plant.

261. The grain is, as you will see, about the size of a large pea, and there are from two to three hundred of these grains upon the ear, or cob. In my treatise I have shown, that, in America, all the hogs and pigs, all the poultry of every sort, the greater part of the oxen, and a considerable part of the sheep, are fatted upon this corn; that it is the best food for horses; and that, when ground and dressed in various ways, it is used in bread, in puddings, and in several other ways, in families; and that, in short, it is the real staff of life, in all the countries where it is the common culture, and where the climate is hot. When used for poultry, the grain is rubbed off the cob. Horses, sheep, and pigs, bite the grain off, and leave the cob; but horned cattle eat cob and all.

262. I am to speak of it to you, however, only as a thing to make you some bacon, for which use it surpasses all other grain whatsoever. When the grain is in the whole ear, it is called corn in the ear; when it is rubbed off the cob, it is called shelled corn. Now, observe, ten bushels of shelled corn are equal, in the fatting of a pig, to fifteen bushels of barley; and fifteen bushels of barley, if properly ground and managed, will make a pig of ten score, if he be not too poor when you begin to fat him. Observe, that every body who has been in America knows, that the finest hogs in the world are fatted in that country; and no man ever saw a hog fatted in that country in any other way than tossing the ears of corn over to

him in the sty, leaving him to bite it off the ear, and deal with it according to his pleasure. The finest and solidest bacon in the world is produced in this way.

263. Now, then, I know, that a bushel of shelled corn may be grown upon one single rod of ground sixteen feet and a half each way; I have grown more than that this last summer; and any of you may do the same if you will strictly follow the instructions which I am now to about give you.

1. Late in March (I am doing it now), or in the first fortnight of April, dig your ground up very deep, and let it lie rough till between the seventh and fifteenth of May.

2. Then (in dry weather if possible) dig up the ground again, and make it smooth at top. Draw drills with a line two feet apart, just as you do drills for peas; rub the grains off the cob; put a little very rotten and fine manure along the bottom of the drill; lay the grains along upon that, six inches apart; cover the grain over with fine earth, so that there be about an inch and a half on top of the grain; pat the earth down a little with the back of a hoe to make it lie solid on the grain.

3. If there be any danger of slugs, you must kill them before the corn comes up if possible: and the best way to do this is to put a little hot lime in a bag, and go very early in the morning, and shake the bag all round the edges of the ground and over the ground. Doing this three or four times very early in a dewy morning, or just after a shower, will destroy all slugs; and this ought to be done for all other crops as well as for that of corn.

4. When the corn comes up, you must take care to keep all birds off till it is two or three inches high; for the spear is so sweet, that the birds of all sorts are very apt to peck it off, particularly the doves and the larks and pigeons. As soon as it is fairly above ground, give the whole of the ground (in dry weather) a flat hoeing, and be sure to move all the ground close round the plants. When the weeds begin to appear again, give the ground another hoeing, but always in dry weather. When the plants get to be about a foot high, or a little more, dig the ground between the rows, and work the earth up a little against the stems of the plants.

5. About the middle of August you will see the tassel springing out of the middle of the plant, and the ears coming out of the sides. If weeds appear in the ground, hoe it again to kill the weeds, so that the ground may be always kept clean. About the middle of September you will find the grains of the ears to be full of milk, just in the state that the ears were at Jerusalem, when the disciples cropt them to eat. From this milky state, they, like the grains of wheat, grow hard; and as soon as the grains begin to be hard, you should cut off the tops of the corn and the long flaggy leaves, and leave the ears to ripen upon the stalk or stem. If it be a warm summer, they will be fit to harvest by the last of October; but it does not signify if they remain out until the middle of November, or even later. The longer they stay out, the harder the grain will be.

6. Each ear is covered in a very curious manner with a husk. The best way for you will be, when you

gather in your crop to strip off the husks, to tie the
ears in bunches of six to eight or ten, and to hang
them up to nails in the walls, or against the beams
of your house; for there is so much moisture in the
cob that the ears are apt to heat if put together in
great parcels. The room in which I write in London
is now hung all round with bunches of this corn.
The bunches may be hung up in a shed or stable for
a while, and when perfectly dry, they may be put
into bags.

7. Now as to the mode of using the corn: if for
poultry, you must rub the grains off the cob; but if
for pigs, give them the whole ears. You will find some
of the ears in which the grain is still soft. Give these
to your pig first; and keep the hardest to the last.
You will soon see how much the pig will require in
a day, because pigs, more decent than many rich
men, never eat any more than is necessary to them.
You will thus have a pig; you will have two flitches
of bacon, two pig's cheeks, one set of souse, two
griskins, two spare-ribs, from both which, I trust in
God, you will keep the jaws of the Methodist parson;
and if, while you are drinking a mug of your own
ale, after having dined upon one of these, you drink
my health, you may be sure that it will give you more
merit in the sight of God as well as of man, than
you would acquire by groaning the soul out of your
body in responses to the blasphemous cant of the
sleek-headed Methodist thief that would persuade
you to live upon potatoes.

264. You must be quite sensible that I cannot
have any motive but your good in giving you this

advice, other than the delight which I take, and the pleasure which I derive, from doing that good. You are all personally unknown to me: in all human probability not one man in a thousand will ever see me. You have no more power to show your gratitude to me than you have to cause me to live for a hundred years. I do not desire that you should deem this a favour received from me. The thing is worth your trying, at any rate.

265. The corn is off by the middle of November. The ground should then be well manured, and deeply dug, and planted with EARLY YORK, or EARLY DWARF CABBAGES, which will be loaved in the latter end of April, and may be either sold or given to pigs, or cows, before the time to plant the corn again. Thus you have two very large crops on the same ground in the same year.

INSTRUCTIONS

FOR USING THE

MEAL & FLOUR OF INDIAN CORN,

PUBLISHED BY THE LATE MRS. COBBETT IN 1846

INDIAN CORN BREAD

DURING a residence in America, from 1792 to 1800, it was not my habit to have bread of any kind made at home in large quantities, but only in small batches of two or three loaves at a time. Such was the custom in that country at the time, and on going there again many years afterwards, I found the same custom still prevailing. This was occasioned by its being so much the fashion, among all classes of persons, to eat cakes, instead of bread for breakfast and tea. They used to make a great variety, not only of the sweet kinds, but such as the Buckwheat-cake and the Rye-cake (see receipts), which are always eaten hot, and which, so far as I could observe, few strangers fail to admire. During the time, however, that Mr. Cobbett was cultivating the Indian Corn in 1828 and 1829, I had bread made of it in my own house at Kensington after the usual manner of our other household bread, and baked in a brick oven. Indian Corn will not make bread of itself, at least

not what we call household bread, that is to say,
large loaves to be eaten cold; it is not adhesive
enough; by itself, it is best made into flat cakes to
eat hot. Having measured the quantities of both
sorts of flour, and allowed one-third of Indian corn
flour or meal to two-thirds of wheat flour, you must
first scald the corn flour as follows: pour over it as
much boiling water as you think will touch every
particle of it, and make it swell, but without making
it really wet, stir it about well, that it may all imbibe
the moisture, then put it out in little heaps on a
cloth to cool, and when it is sufficiently so for you
to hold it in your hand, rub it smoothly and thoroughly into the wheat flour; this done, set it to rise,
and proceed as you have been used to make your
bread.

Bread is made in different ways in this country,
but the way I have been most accustomed to is generally, I believe, practised in the Southern counties,
and I give it here, extracted from Mr. Cobbett's
" Cottage Economy."

" Suppose the quantity be a bushel of flour. Put
" it into the baking trough, and make a deep hole
" in the middle of it; mix a pint of good yeast in a
" pint of lukewarm water, or half milk and half
" water, and pour it into the hole; then take a spoon
" and work it round the outside of this body of
" moisture so as to bring into this body, by degrees,
" flour enough to make it form a thin batter, which
" you must stir about well for a minute or two; then
" take a handful of flour and scatter it thinly over
" the head of this batter, so as to hide it. Cover the

" hole over with a cloth to keep it warm; and this
" covering, as well as the situation of the trough,
" as to distance from the fire, must depend on the
" nature of the place and state of the weather as to
" heat and cold. When the batter has risen enough,
" cracks will appear in the flour that you covered it
" with, and then you begin to work the whole into
" dough, thus: first, scatter in half a pound of salt,
" then begin round the hole containing the batter,
" working the flour into the batter with your hand,
" and pouring in, as it is wanted to make the flour
" mix with the batter, soft water, or milk with water,
" or milk alone, a little warm. When you have got it
" sufficiently moist, you must knead it well, and this
" done, let the dough stand covered over, and in a
" warm place, about fifteen or twenty minutes, or
" till your oven is ready."

I do not say that a greater proportion of Indian
corn meal may not be used, but I do not recollect that
I ever tried it.

BREAD AS MADE IN ITALY

Use half Polenta (which see) and half wheat flour.
This bread is moister than that made after the above
receipt; and more so, perhaps, than English people
generally would like; and being moist, should be
baked in tins, or in flat cakes, like what Hampshire
people call oven cakes.

ASH-CAKE

This is as commonly made in Italy as in America;
and is bread baked in the ashes, and is made wholly

of Indian meal, and not fermented. Scald the meal, put it in little heaps to cool, then mix it with more water (warm), into dough, and mould it into flat cakes rather larger than a breakfast saucer. These are baked as follows: open a place in the side of a wood fire on the hearth, and having put in the cakes, each between two cabbage leaves, lay them on the hot hearth, sprinkle some ashes lightly over first, then put hot coals on the top, and if these appear to cool fast, remove them from time to time, and replace them with hotter coals from the fire. As the time required for cooking these will depend upon the degree of heat in the coals, I must leave it to the experience of the bread maker to decide upon that point for herself. I have been told that in Devonshire and in some other parts of the country, it is or was the custom, in cottages and farm houses, to bake bread on the hearth, large loaves as well as cakes, covering the loaf close with an iron, or brown earthenware vessel.

This way of making bread would surely be very serviceable to the poor Irish, who have got their turf fires, and also sweet buttermilk to wet the meal up with; which would make the cakes better than water. I wish they could be taught to try them.—(See " Buttermilk Pudding.")

JOHNNY CAKE

This is made in the same manner as Ash-cake, and baked before the fire, first on one side, and then on the other.

PONE-CAKE

Made the same as the Ash-cake, and baked in a utensil much in use in America, and also in some parts of England: a round kettle standing on legs, with a lid fitting down, so that you may put hot coals on the top, as well as under it. It stands by the side of a wood fire, and is a most useful article of domestic furniture in houses where there is a wood fire on the hearth. I am surprised that it is not in more general use with us, for with the experience I had of it in America, I do not think I could live in the country without one. I baked, not only bread, but cakes large and small, puddings, custards, apples and pears in it. It stands apart in any corner of the hearth, and is altogether one of the most convenient things I know of. This and the girdle should be in every labourer's cottage.

MUSH

I do not know whether it be the prevailing custom, but I have always heard the flour of Indian corn called meal, whether in a coarse state or fine, but for Mush it should be coarse. This Mush is, in fact, Porridge, and like the latter, is made thick or thin, as you like, but I make it as follows: having five pints of water or milk, whichever you prefer, boiling fast on the fire, put in a small tea-spoonful of salt, and while you keep stirring with your right hand, drop gently from your left hand, and by degrees, one pound of meal: let it boil twenty minutes, stirring well all the time or it will be lumpy; a stick is best.

If the meal be fine, then it is better to wet it up with cold water into a smooth stiff batter first, and stir that into the boiling water; but you must stir it all the time it is boiling, and in this case as well as in the other. The above is eaten in America, in the same way that porridge is eaten in some parts of England, namely, having some Mush in a plate, and a cup of milk by the side, you take a little of each in your spoon, and it is really very nice. This makes a good breakfast for grown-up persons; and I see that Mr. Cobbett, in his corn book, says that about three pounds of meal will make Mush enough for the breakfasts of ten grown persons. For children it is excellent food, both satisfying and nourishing.

A poor woman would soon find the benefit to her children, if she could give them as much Mush and milk as they could eat. To my taste, it is much more palatable than oatmeal porridge, and I think there can be no doubt of its being more wholesome. Mush would, I think, be found upon trial to be an acquisition in families of the middle class, as well as the poor. It is good eaten like Hasty Pudding, with butter, sugar, treacle, preserves, or fresh fruit, first stewed or baked, such as apples, pears, gooseberries, currants, &c., &c. When cold, Mush is solid like ground rice pudding, and may be cut in slices and toasted before the fire, or just browned on the grid-iron, and eaten with roast meat; or after the meat with butter and sugar or preserves. In fact, a tolerably skilful housekeeper would soon discover that it is capable of being converted into a variety of nice dishes—(See " Polenta.")

POLENTA

The best thing to prepare this in is a three-legged iron pot, hung over the fire. Let the fire be hot, and also blazing, if possible. To a quart of water, when it boils, put in a little salt, then add twelve ounces of meal, but be careful to do it in the following manner: while the water is boiling, stir in half the meal first, but be sure to stir quickly all the time or it may be lumpy; then you may put in the remainder at once, but keep stirring constantly. When it has been on the fire a quarter of an hour, cease to stir, take the pot off the fire, and set it on the floor for two minutes, then put it on the fire again, and you will see the Polenta first rise in a great puff, then break and fall. As soon as you perceive this, take it off the fire, and turn it out into a dish; it ought to come out quite clean, not leaving a particle adhering to the pot, else there has been some fault in the boiling. It is stirred with a long stick, thicker at one end than the other. Of this the Italians make an endless variety of dishes, some of which are the following.

The most simple mode of dressing the Polenta is thus: pour it from the boiling into a bowl; when cold, turn it out; take a coarse thread in your two hands, put it on the side of the Polenta away from you, draw the thread towards you, and you will find that it cuts a clean slice of Polenta off; continue till you have cut it all into slices, and then you may dress them in different ways. The commonest is to cut the slices thick and brown them on a grid-iron, as I have directed for cold Mush.

ANOTHER WAY

Butter a dish, lay thin slices in, then a layer of force-meat, or bits of bacon, or minced onions browned, or Pesto (which see), or hard boiled eggs, chopped, or oysters chopped; in the case of the last two, put little bits of butter, pepper and salt, then a layer of Polenta, and so on till you have enough, then bake it. This dish may be variously, and highly, flavoured, according to taste.

POLENTA WITH CHEESE

Put spoonfuls of Polenta all over a dish, grate cheese over, then stew little bits of butter, pepper, and salt, then another layer of Polenta, and so on till you have enough, and brown it before the fire or not, as you choose. In place of butter you may put gravy, either of mushrooms or meat.

PESTO

Take bazil, majoram, parsley, onion or garlic, pepper, salt, and a bit of cheese (it does not signify how old and dry), and beat all well in a mortar; mix it with boiling water into a smooth paste. In Italy they add fine oil, and you may add fresh butter. Then thin the paste with boiling water to the consistence of melted butter, boil it for a minute, and dress it with Polenta, as in the last receipt. This is a savoury dish, and the mixture is much used in maîgre dishes.

MINESTRA

This, in Italy, I believe, means anything of the liquid kind made very thick, thus: drop meal into boiling soup, enough for this purpose, keep stirring till it is quite cooked, about twenty minutes.

ZUPPA

The same as the last, but thin, with very little meal in it.

SUPPAWN

An American dish, but the same thing, in point of fact, as the two last. It is very commonly the breakfast and supper of labouring men in the country, and is made by thickening broth with Indian meal, and is really very good. Surely it would be very good in soup which is made for the poor.

Mr. COBBETT used sometimes to have his Beef Tea thickened with the meal.

SAMP

This is the corn when rubbed off the cob just cracked (not ground), and the skins rubbed off. It is very good for poultry, and it is used for making soup, like split peas. The Americans boil it with pork. They call the corn when cracked Samp, and I believe that they call the dish, which it helps to compose, Samp also. I never had it in my house that I recollect, but I have eaten and liked it; and I think that the coarser parts of beef, neck of mutton, and even the old cock which Scotch people dress up with leeks, might be cooked with Samp to advantage. Samp requires very long boiling, or it will not be good at all; but really well done, I think would be very good as a change, in a labourer's family.

HOMMONY

This is a dish that I really am not practically acquainted with, so that I can only give a receipt which has been given to me by persons who have it constantly at their table in England. Boil one-third of a pound of meal in enough water to cover it, for twenty minutes, or until nearly all the water is wasted; it must be like thick paste. Put a piece of butter the size of a walnut into a vegetable dish, pour in the Hommony, and serve it like mashed turnips. Dip your spoon in the middle, when you help it. There may be other ways of cooking Hommony, but I am not prepared to give directions for them. I understand that in some parts of America, what they call Hommony is made of the cracked corn; and if so, it must be something of the same kind as our peaspudding, but not boiled in a cloth. This would be worth trying, for I do not see why it should not make a pudding like peas, and be very good; at all events it would make a variety.

PLUM PUDDING

To one pound of meal, add half a pound of shred suet, and what currants, raisins, and spices you choose: mix the whole well together with a pint of water, and boil the pudding in a cloth three hours. This is the receipt I have followed, and made it more or less rich as required.

SUET PUDDING

Put half the quantity of suet that you have of meal, not very finely chopped, and a little salt: boil in a cloth the length of time that you would a wheat flour suet pudding. Wet it up with water, to the same stiffness that you would if made with wheat flour. Apples chopped make this a nice pudding.

DUMPLINGS

These are made of meal wet up with water or milk, and boiled in a cloth, to eat with bacon or pork; and I. think would not be objected to in the farm houses of those counties where Dumplings prevail.

BUTTERMILK PUDDING

Sweet buttermilk is an exceedingly nice and wholesome thing, and it is a mistake to suppose that buttermilk is, for the greater part, sour; the Irish at all events know better, and I wish that some of the ladies of that country would promote the use of Indian meal as an accompaniment to it, first in their own houses to set the example, and then in teaching the poor to cook and to eat it. In the country I frequently had this pudding, and I recommend a trial of it. Mix the meal into a stiffish batter with butter-milk and a little salt, tie it in a cloth, and boil as long as you do other batter puddings. We eat this with roast meat gravy, or with butter and sugar.

A NICE PUDDING

Made like Ground Rice. Mix two ounces of meal smooth with half a pint of cold milk, and pour over it three half pints of new milk scalded, stir it over the fire till it thickens: let it cool, then add four eggs well beaten, sugar, nutmeg, and a spoonful of orange-flower water, beat it, and pour it into a dish with a a paste border. Bake it half an hour. Or, adding two more eggs, this may be boiled in a basin, an hour. This may of course be flavoured in any way you choose.

YORKSHIRE PUDDING

Make this the same as the last receipt, but use a little more meal to the same quantity of milk, and only three eggs. Do not turn the pudding. No sugar.

GRIDDLE CAKES

Various cakes are made in America bearing this name, and they take it from the implement used in their cookery, which implement is known in some of our Northern Counties as the Girdle and Pie-Klit, and many are, I believe, manufactured at Birmingham: but I have never heard of its being used in the South of England. The process is the

same as that of making Pancakes, or of baking
Crumpets and Muffins: in London the two latter
are all baked on a hot plate on a stove, and this
answers every purpose of a girdle; but in labourers'
cottages, and houses where there is no hot plate, I
think the girdle would be found a useful appendage.
It is a flat round iron, of diameter sufficient to cook
a good-sized pancake, with a handle to hang it on
the crane, and some of them have three short legs
to stand on, when it is put aside. In cooking pancakes, a good piece of butter is put into the pan;
but when baking cakes on the girdle or hot-plate,
use a little piece of suet on a fork, or a bit of butter in
a thin rag, and gently rub the girdle all over, in order
that the cakes may not burn, and yet not be greasy.

BUCKWHEAT CAKES

I give this receipt because the cakes are very nice;
and I should suppose the grain may be imported at
a very cheap rate, therefore it would be an acquisition to us. But there needs great particularity in the
dressing, for Buckwheat flour is apt to be gritty, even
in America, where it is so much eaten in these cakes.
They are made as follows:—Mix a pound of the
Buckwheat flour into a batter with warm water,
rather thicker or stiffer than for pancakes, in a deep
pan or jar, and stir in from one to two table-spoon-
fuls of yeast, according to its quality, cover it over,
and put it near the fire to rise; in three hours it will
be ready to bake. The girdle or hot plate being quite
clean, and as hot as you would make the frying-pan

for pancakes, take a piece of suet on a fork, and rub it well over, then dip a tea-cup into the batter, take it out full, and pour it on the middle of the hot plate, when it will spread out to the size of a breakfast plate: when it is sufficiently done on one side, turn it with a knife, as you would a pancake. When one cake is done, you must rub the girdle again with suet, or the next cake will burn. When you have poured the second cake on the girdle, rub some butter over the first one, and put it before the fire to wait for the rest, and so on to the end. Be not alarmed at their colour, which is always of a dark hue, for a pile of hot Buckwheat cakes on a breakfast table in the country, of a cold frosty morning, is not to be despised. The French make bread of this flour, but it is heavy, and merits its name:—Pain Noir, black bread.

CORN CAKES

Scald the meal in the way directed for bread (which see), and then mix a pint of meal and two eggs, yolks and whites, with warm milk, into a batter, rather thicker than pancakes, and bake the cakes as directed for Buckwheat cakes. We have made them without any wheat flour, but a table-spoonful may be added, and is an improvement, because the meal is so crumbly, if I may so express it. A very little salt is also put in, and remember, that, as in the case of all batter, it must be well beaten to be light. This may stand two or three hours, but then it must be beaten well, just before baking.

ANOTHER WAY

Use half wheat flour and half meal, two eggs to a quart, some salt, and make it into a thin batter with warm milk: bake on a girdle the same as above.

MUFFINS

Mix a quart of meal, previously scalded, into a thick batter, thicker than the last receipt, with hot water, and two eggs, and some salt, beat it very well, and then put in a table-spoonful of yeast, beat well again, and let it rise before, or by the side of the fire, for three hours. You must have Muffin rings for these, and bake them on a girdle or hot plate, as in the last two receipts.

RYE CAKES

Mix a quart of Rye flour, three eggs, yolks and whites, and salt, with milk, into a batter, rather thicker than for pancakes, and bake on the girdle, in the same way as buckwheat. More eggs may be used. These depend upon the delicacy of the flour for their goodness, and I have never seen it dressed sufficiently fine in this country; but it may be to be had nevertheless, and would make an agreeable variety.

BREAKFAST CAKES

Mix one pound of meal with three eggs and enough warm milk to make it into a very stiff batter, or, rather, very wet dough, as you mix a cake to bake in a tin: put it on a tin into a hot oven (it should rise quickly or it will spread too much), and bake it twenty minutes; cut it as you do Sally Lunns, butter it, and serve it hot.

FRENCH PIE

The above mixture makes a very nice covering for cold meat, in a pie dish, as some persons use mashed potatoes. The meat being in slices, peppered and salted, with a little gravy, or a bit of butter rolled in flour, lay the paste over the top, and bake it: the crust will be done enough by the time the meat is heated through.

A NICE CAKE

Break three eggs into a pan, put to them six ounces of meal, four ounces of sugar, the grated peel of a lemon, the yolks of five and the whites of three eggs, and a table-spoonful of orange-flower water, beat it well for twenty minutes, pour it into moulds, and bake the cakes three quarters of an hour, of a light brown colour.—This is the same as rice cake.

INDIAN CORN BISCUITS

To half a pound of butter six ounces of pounded sugar and three eggs; when these are well mixed, beat them up with three quarters of a pound of meal, some nutmeg and carraway seeds: bake on little tins.

ANOTHER WAY

Four ounces of butter to three quarters of a pound of meal, four ounces of sifted sugar, and nearly an ounce of carraway seeds; make it into a paste with three eggs, roll it out thin, and cut it in shapes, as you like.

N.B. I have not made a constant practice of it myself, but I think that it is best to scald the Indian flour or meal, as directed in the receipt for bread, before using it in any species of cookery.

THE END

INDEX